Understand Your Heart

The Heart-brain controls more than we thought

By Steve Preston

1st Edition

© Copyright 2017, Steve Preston
All rights reserved.

Contents

Understand Your Heart 1
Introduction 7
Triune Entity 14
Ancient Book of Job 20
Moses Descriptions 23
Jeremiah Writings 27
David's View 29
Solomon's View 31
Other Prophets 33
God's Heart-Brain Descriptions 35
Paul's Writings 38
Distributed Brains 42
Ant Brains 44
Heart Brain Biogenetics 48
Heart and Head-Brain Communication 51
Body Hormone Controller 55
E-M Messaging to the Brain 57
Subsonic Brain Manipulation 65
Infratonic Stimulation 67
Transcranial Magnetic Stimulation (TMS) 70
Heart-Brain E-M Emitter 72
Healing with Hands 78

Faith-Healing	82
How Do You Get Faith?	85
Heart and Emotions	87
Heart Consciousness	88
Heart Brain Learning	90
Heart-Brain Control	92
Communicating Heart Brain	96
Soul Focused Heart-Brain	98
Heart-Brain Sight	103
Intuitive Heart-brain	106
Hardened Heart-brain	108
Heart Replacement Caution	117
Reincarnated Heart	123
Long Term Memory	129
Michael Newton's Soul Journey	135
Circumcised Heart	138
Spirit Wisdom	146
Chakra & Heart Brain	150
Soul Enhancement Levels	154
Abraham Maslow	158
Anthropics	160
More on Soul Control	164
Heart-Brain Globalization	166
Music and the Heart-Brain	168

Expanded Review 170
About the Author 174

Introduction

There is a problem with people trying to understand their life, their interactions with those around them, how they think, how they can control their destiny, how they view the world, and even how they love. Quite simply, they ignore or don't know how to interact with a very important part of themselves and, possibly, the most important part.

They do not listen to or use their heart.

I'm not trying to be poetic about our heart or talking about a cute little cupped hand making a squashed circle symbol.

I'm talking about one of the most fascinating and important parts of our entire being. We are so caught up in the flowery "heart/love remarks we think they were the basis for the 513 Bible verses describing the heart and it importance to our lives. Scientists now have a different understanding as many of the descriptions in the Bible have been tested in the laboratory and the heart has been found to be so much more important than we had ever imagined. The Bible described the heart as this master of our body and soul, we are finding out about <u>its independent brain</u> and how most of the messaging between our heart and brain initiate in our heart-brain. The Bible continued by stating the heart could conquer and destroy our lives, loves, and even our redemption. Researchers ran test after test and the more they tested the more they found the heart brain was responsible for communication and control inside us and outside our bodies. Other factors described in the Bible are somewhat harder to clinically test as the Bible claims the Heart interfaces and influences a portion of our being called our soul and interfaces with another component of our being called our spirit but even those things are becoming more evident. Here is the part that is very important. There are not 513 verses describing the wonders of the heart just to make some grandiose imagery of a "caring" person, or a "loving" person, or a "hateful" person. The Bible is not silly poetry and it is not providing misinformation as some would suggest when it describes great power in the capability of our heart as if it had its own brain that was three times as fast and had 5000 times the Electromagnetic [E-M] communication [radio wave] distance of our head-brain outside the body. Ways to focus that energy seem to be very effective and we find hints of how to use the energy are found in the Bible.

Over the past 50 years, heart studies have investigated multiple characteristics and each time, the heart is shown to be in control of more and more of our lives. From scientific experimentation and discovery, we now know, for instance, there are two brains in our bodies; one brain is in the head and a second, faster brain, is integrated in our Heart. Our hearts think independently and react on just about everything we do, say, or ponder. They initiate conversation with heart-brains of those near us. They control many of the hormones that regulate all types of things in our bodies including stress, learning, and love. While all of these capabilities are amazing, studies show a long-term memory in the Heart-brain that is proving to be a very important part of its total function. The following timeline provides a general description of the work done in this area.

1800BC- In ancient Canaan, researcher Job, also known as Jobab, described the Heart as being guilty and godless on its own and full of resentment. He or a scribe wrote his findings in the "Book of Job" and it was later inserted as an important book of the Bible collection. He challenged that this master heart would yearn for answers and that it was <u>intimately communicating with your mind.</u>

1500BC Coming out of Egypt, a researcher named Moses came along and described the Heart as the major controller of our bodies and actions. He also initiated the idea of a "hardened heart" that would cause irreparable damage to our lives as it fostered an impenetrable, egocentric view of the world. He also was first to coin the phase "circumcision of the Heart" describing its reactions outside the body.

*1000BC- In the land of Israel, researchers and Kings David and Solomon expanded the understanding by describing the heart as controlling or affecting Wisdom, Perversity, Lust, Laughter, Joy, and Anxiety in our lives. They also expanded the understanding of the interface between our heart and the entity known as the **<u>Spirit</u>** as an integral part of our lives.*

600BC- Researchers Isaiah, Ezra, Ezekiel, Hosea, and Jeremiah took up the study and again and indicated the heart and mind communicated as if it had its own brain and continued descriptions of circumcision of the heart to make one's life better.

30AD- Jesus, the incarnate God, described our hearts as the major controller of our bodies and hinted that it can control reality and allow performance of miraculous feats when the heart became more strongly linked with the **Soul** component of our being. He called this linking "faith". He also described a linking between the Heart and **Spirit** portion of our being that allowed our lives to be closer to God. To make things clear he called this "faith in God".

***70AD*-** Researcher and apostle Paul described the Heart as the most important part of us; <u>linking our spirit, soul, and mind together</u> to allow healing of sickness and other miraculous deeds.

***1000-1800s*-** There was a time when the Bible was not considered an important scientific work. During that time, the Heart was inappropriately described as a simple blood pumping organ. The idea was accepted by consensus throughout the world. This would greatly slow development of Heart-brain studies around the world.

***1960s*-**Pioneers in the re-investigations of the Heart-brain, John and Beatrice Lacey rediscovered that the heart initiated communication with the Brain that effected how we perceive the world.

***1970s*-** Researchers developed what was called Infratonic devices that would flood the brain with subsonic cues to reduce stress similar to the Heart messaging not knowing the Heart did the same thing.

***1983*-**As hormone transfer was one of the main internal communication methods, the heart was reclassified as part of the hormonal system showing its importance was well beyond blood pumping.

***1991*-**Reasearcher Dr. Childre set up the HearthMath research Institute to help individuals, organizations and the global community <u>incorporate the heart's intelligence into their day-to-day experience of life.</u>

***1992*-** A researcher in the Medical Unit in Pavia, Italy, Luciano Bernardi, published data on Heart Rate Variability [HRV] messaging and indication.

1994 Neurophysiologist named Dr. Armour made strides in defining the heart-brain.

1995- *Researchers Rein and Atkinson indicated that Heart brain massaging was used to calm the head brain.*

1997- *Rollin Mcraty and HeartMath found hormonal messaging controls the heart greatly influenced our actions and coherence-building techniques lowered disruptive Heart Rate Variations HRV.* Remember this coherence word as it links our heart-brain to our head-brain and reduces stress.

1999-Mcraty and the HeartMath team showed two-way dialog has the heart initiating most messages.

2000-Researchers Murphy, Thomspon, and Mcraty described how the heart-brain memory affect transplant patients' capability.

2000s- Development of an E-M emitter almost identical to the E-M communication of the Heart was developed and continues to have great success treating pain, stress, and sickness. Called the Transcranial Magnetic Stimulator, no one knew they were simulating the exact actions of the Heart brain.

2002- Quantum Intech was formed to develop technologies based on HeartMath research on heart-brain-body communications, enabling people to better harness the intelligence of the heart to improve their performance and health.

2011-Researchers Taggart with Critchley found through neuroimaging that a specific set of cortical and subcortical brain regions were involved in cardiac control or arrhythmogenesis.

2013- From additional work at the HeartMath Institute, it appears that intra-person Heart communications may extend to some world level phenomenon.

2015-People are finally beginning to take a second look at what was said thousands of years ago about our all-important Heart and its brain.

When researchers finally got away from thinking the brain was a simple blood pump and tried to find out how it was able to do some of the things it was doing, they looked for synapse and neurons or Ganglia. They were not disappointed and found brain functional elements throughout our heart as shown next. This distributed brain characteristic seems to give the heart-brain the effect of providing faster, free-er access, computation than head-brain computation. According to the research Institute HeartMath the dots in the following diagram are ganglia clusters or small brain masses. [Ganglia positions are noted as "brains".]

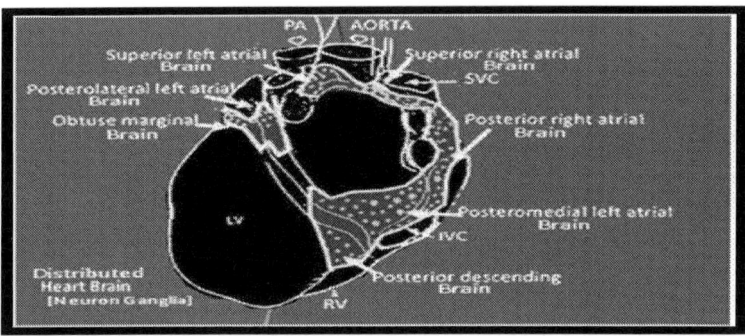

Researcher "Moses", in the Bible, described many of the capabilities of the heart-brain thousands of years ago and we should not ignore what was said. Let's just look in Genesis for a second.

Genesis 6:5 The LORD saw - every inclination of the thoughts of the human heart was only evil all the time.

I know that verse is not saying anything good, but the thing we need to recognize is Moses wrote that the heart was a thinking tool. Like our head-brain, he told us the heart-brain is easily misguided such that it can turn to evil. He wasn't stupid, he didn't have misguided information about the heart, and he was not spouting poetic verse. He simply knew something we are only now learning. This book will examine some of the Biblical texts [not all 513] to look for clues along with substantial scientific research investigating our heart brain to understand what Moses knew 35 hundred years ago. King Solomon provided us with the meat of this entire book 3000 years ago. He wrote the following;

Proverbs 4:23 *Above all else, guard your heart, for everything you do flows from it.*

No! Blood is not EVERYTHING in the world so don't go believing Solomon was saying nothing exists without blood. He was not talking about that part of the heart at all. The following diagram shows what he was saying.

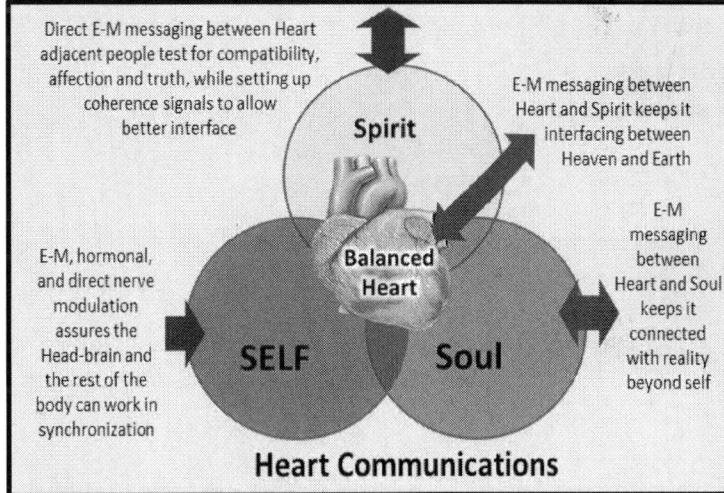

First thing you see is three circles representing our triune entity. Made in the triune image of God, we are not simply a living organism like a blade of grass. Instead we portray something physicists call cognition.

Triune Entity

To understand the Heart-brain messaging, we must recognize our complete self. We are cognizant of ourselves [by our Head-Brain], our reality [by something called the soul], and our linked universe [by something called our soul]. Scientists tell us without these levels of cognition, nothing would exist. We are finding out that we cannot interface with the soul or spirt part of us without the almost unbelievable capabilities of our heart. Another way of showing our triune entity is next. The heart communicates with all three.

As we dig into the capabilities of the Heart, we will have to look beyond our "self" portion, so some of this may sound a little bizarre if you have not had instruction on the three entities of our being, so I will try to expand your awareness as slowly as I can. First let's look for verification in an unlikely location.

Triune Egyptian Entity and Heart

The Egyptians called our triune entity the Ba, Ka, and Shut. **Ka-Self**-The person was the *Ka [self- depicted as a bird with human head and long arms]*. Exactly like the Jewish description, the Ka was the portion of a person that dies. The following comes from the "Book of the Dead".

[God] knows **the names of your ka,** *the* **aspect of your soul** *that abides in the ground: Nourishing ka, ka of food, lordly ka, ka the ever-present helper, ka which is a pair of kas begetting more kas, healthy ka, sparkling ka, victorious ka, ka the strong, ka that strengthens the sun each day to <u>rise from the world of the dead</u>, ka of shining resurrection, powerful ka, effective ka.* Like the Jewish depiction the Ka, self needs food, health, procreates and dies to be resurrected in a distant time.

Ba-Spirit-The spirit was the *ba, sahu [depicted as a flying eagle with the head of a man]*. Just like the Jewish description, the Ba connected our being to the spirit world of God. The following is from "Egyptian Book of the Dead".

<u>*I go round about heaven and sail in the presence of [God]*</u> *Ra, I look upon all the beings who have knowledge. Hail, Ra, I who goes round about in the sky, I say, O Osiris in truth, that I am the Sahu of the god, and I beseech you <u>not to let me be driven away,</u>* Later we will see the concept of

hardening the Heart is actually allowing the spirit to be driven away so that it cannot communicate with the Heart.

"[God] knows the names of your ba, the form in which you travel our world - the sun. Ba pure of body, health-embodying ba, ba bright and unharmed, ba of magic, ba who causes himself to appear, male ba, ba whose warm energy encourages copulating." Similar to Jewish depictions, the Ba/spirit is the pure part of our existence who knows God and is needed if we wish to communicate with God.

Shut-Soul-and the soul was the *Shut, Shadow [depicted as smoke in a human form].* Exactly like the Jewish descriptions, the shadow/soul was not bound to the grave and could go where the body could not. The following comes from the "Papyrus of Nu".

O *mighty One, when he is adored, great one among* bas, *greatly respected* ba [Spirit] *inspiring the gods with awe when he has appeared on his great throne: then may he prepare the path, justified, his* ba, <u>and his Shut</u>, *may they be well provided for.* In this depiction, just like the Jewish depiction, the Shut separates from the "spirit" after death and is reunited after a resurrection in the future. The following shows one of the depictions of our triune existence showing the Ba, Ka, and Shut.

Egyptian Heart

The Egyptians correctly believed that the heart, rather than the brain, was the source of human wisdom, as well as emotions, memory, the soul and the personality itself. To them, it was through the heart that God spoke, giving ancient Egyptians knowledge of God and God's will. The heart was considered the most important of the body's organs. While they seemed to understand parts about the heart we are just now learning, the heart's role in blood circulation was not precisely understood.

Heart Communication- It was believed that the "heart communications" or "metu", connected all parts of the triune body. Unlike what we know today, the Egyptians thought the main head brain function was to pass mucus to the nose, so it was one of the organs that was discarded during mummification. Sometimes I think my head-brain is designed just to pass mucus as well. Here is one saying from the "Book of the Dead".

17

O my heart which I had upon earth, do not rise up against me as a witness in the presence of the lord of things; do not speak against me concerning what I have done, do not bring up anything against me in the presence of the great god of the west..."

Sigmund Freud and the Triune person

In modern times, people have struggled with the definitions of the triune existence of humans and the involvement of the Heart. One man came along and tried to develop a new concept from what he saw, but he didn't follow the details of the Early Jewish people and the Egyptians so he got things mixed up a bit and by this time the Heart had been pushed aside as a simple blood pump. This man was Sigmund Freud and he tried to redefine the elements of life into his own concept to try to make it seem that this reality could hold the essence of the three dimensions of a triune existence. Sigmund Freud tried to connect the differences in characterizing a person without using ancient religion to guide him. He came close, but he missed important aspects.

In Freud's model of the psyche, the ID (instinctive unconscious, soul), the Ego (organized, conscious self), and the Superego (moralizing, not entirely unconscious, spirit) form an interactive framework which work together in the mind. Here is what he had to say.

"One of the fundamental functions of the Ego is Reality Testing – reaching into the real world to see if what is believed to be the case actually proves out – but this does not bear full fruit until the Ego has become Autonomous... substantially set free from inner conflicts between the ID and Superego."

This is sort of backwards from all other descriptions, but one thing he got right. In his description, the EGO, the ID, and the Super-Ego were continuously at war. The depiction below is a common way of showing Freud's depiction. Unfortunately, he did not explain how these three entities were joined.

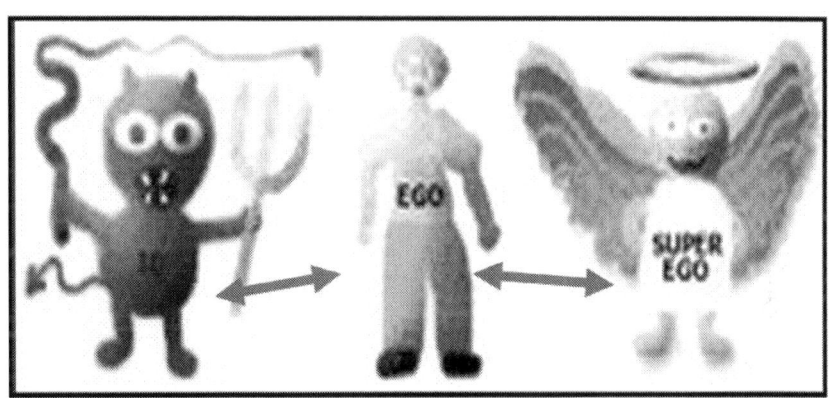

Besides thinking the soul was the evil entity and the self was the uncommitted portion of our being, his lack of knowledge about the heart would not let his understanding go farther. Science had not given him clues he needed. We will certainly look at the wisdom of science in this book, but for those still thinking Moses and Solomon were crazy, we will look at the words of the incarnated God; Jesus or Joshua, depending on how you interpret Greek words. One example is provided below from the Gospel of Matthew.

> ***Matthew 13:15** For this people's heart has become calloused; - Otherwise they might -- <u>understand with their hearts and turn, and I would heal them</u>.'*

Just like Solomon told us, the heart controls everything about your entity. If we quit using our heart to communicate with our soul and spirit, we will be locked away and sick in our "self" and become calloused. The only hope is to use you heart-brain. Then you will be "healed". There is no question the creator of our universe described the heart as having an understanding brain. Now we know he was not lying. Some of you are thinking the stupid heart pump cannot be associated with all this and if it was, how in the world could such a tiny thing have the power to do all of this. By itself it can't, but it is intricately tied to our body and it uses that source for communications. We will get into all that later

In this book, we not only investigate the old descriptions, the modern experimentations, and discoveries about the Heart-brain, but we will also investigate how the heart brain can be so powerful and modern heart-brain stimulation devices being successfully used to communicate with our head-brains, but some of our best information comes from the Biblical stories.

Ancient Book of Job

Some tell us the book "Job" is the oldest book in the collection of book known as the Bible so let's see what was said about the Heart-brain even before Moses' stories were collected and distributed and see if there was a difference as there is no evidence that Job ever met Moses or any of the Jewish colonists that went down to Egypt. No one knows who wrote Job, but we finally have some insights into who Job was and when he was. For some, this might be a surprise. In this history, we find that a Shuhite, a Tamanite, and a Buzite all talk to Job.

Tamanites -were descended for Teman the first son of Eliphaz which would have made him come from Edom and the story would have been some short time after 1800BC.

Shuhites- were descended from Shua Abraham's son by the concubine Keturah. He was an Assyrian and would have come from midway up the Euphrates River. This could be timed around the death of Sarah around 1800 BC.

Buzites- were descended for Buz, one of Abraham's brother's sons and would have lived near the city Nahor which is also now considered Assyrian, but farther north than either of the others who had traveled a long way. As Buz lived around 1850BC, this places Elihu around 1800BC as well.

What about Job? -We are told he lived in the East in the land of Uz as the richest man or [king]. While we don't find a Job, we do find <u>Jobab</u> who was the <u>Great Grandson of Esau</u> and his father would have been Zerah so let's test the timing. Esau was born around 1800BC and 20 years between generations would subtract 40 years which would have <u>Jobab being born around 1760BC</u> which is supported by his friends. Jobab was a contemporary of Joseph and Jacob. While we don't need that information to examine the heart-brain, it is always good to understand which thoughts were first and Job describes heart information known before the time of Moses especially believing Jobab came from Esau the bitten enemy of Jacob who finally killed Esau during the great Jacob-Esau War described in the book "Jasher" and others. The War and killing of Esau left the bitter hatred of the Jews by the Amalekites and others Jobab viewed as friends. Let us review.

***Job 10:13** What is <u>concealed in your heart, and I know that this was in your mind:</u>*

***Job 19:27** How my <u>heart yearns</u> within me!*

***Job 31:33** hiding my <u>guilt in my heart</u>*

***Job 36:13** The <u>godless in heart</u> harbor resentment;*

In Job, we find that the heart-brain can "conceal its desire" and it can "yearn for various things". Our heart can "feel guilt" and "harbor resentment". These are not the actions of a muscle; instead, they describe new information we are only now uncovering. I know, I know you are saying in the back of your mind----*"Its poetry; not a description of some diabolical massaging monster inside our chest like the alien baby that finally erupted from a host as an acid spewing thing while the host person simply died."* Wow! I didn't expect that much resistance. The heart is not an alien, but it is EXTYREMELY important for our existence far beyond pumping blood, controlling the blood pressure levels, pumping adrenalin to support actions in concert with blood pressures and the like. As we continue, understand, the ancient writers were trying to explain existence not evoke a flowery picture.

Moses Descriptions

Here is a partial list of descriptions of the Heart-brain by Moses. It makes me so mad when people just gloss over these sentiments or they try to explain the words as being figurative of poetic. Moses was not a poet, but he was a truth teller and we will find later he was telling the truth about our Heart-Brain. Before we look at these let just let you know right now, Moses, was not his Egyptian name. We find that 25 Egyptian rulers had the name Moses as it simply means "born of". Ka-Moses, the last ruler of the 17th Dynasty, for instance meant "Born of -the spirit Ka" and Medu-Moses, last ruler of the 14th Dynasty, meant "Born of the Arrow/War". The first ruler of the 18th Dynasty was named Ah-Moses, which meant "Born or the moon god Ah". As the Jewish Moses, prince during the 15th dynasty, was born of the Nile, his Egyptian name would have been "Al-Moses". Sorry; I got carried away. Here are some of Al-Moses' descriptions.

Genesis 24:45 *"Before I [servant of Abraham] finished <u>praying in my heart</u>---*

Genesis 34:3 *His [the Philistine King's son] <u>heart was drawn</u> to Dinah daughter of Jacob*

Exodus 4:21 *The LORD said to Moses, - I will <u>harden his [Pharaoh's] heart</u>*

Exodus 7:13 *Pharaoh's heart would not listen to them.*

Exodus 25:2 *"You [Moses] are to receive the offering for me [God] from everyone whose heart prompts them to give.*

Exodus 35:21 *everyone whose heart moved them came and brought an offering.*

Leviticus 19:17 *'Do not hate a fellow Israelite in your heart.*

Deuteronomy 4:9 *--do not forget the things your eyes have seen or let them fade from your heart.*

Deuteronomy 4:29 *seek the LORD your God with all your heart and with all your soul.*

Deuteronomy 6:5 *Love the LORD your God with all your heart and with all your soul and with all your strength.*

Deuteronomy 8:2 *God led you forty years, to humble and test you in order to know what was in your heart*

Deuteronomy 8:5 *Know then in your heart that as a man disciplines his son, so the LORD your God disciplines you.*

Deuteronomy 8:14 *then your heart will become proud and you will forget the LORD*

Deuteronomy 10:12 *fear the LORD your God, to walk in obedience to him, to love him, to serve the LORD your God with all your heart and with all your soul*

Deuteronomy 13:3 *love him [God] with all your heart and with all your soul.*

Deuteronomy 15:10 *Give generously without a grudging heart;*

Deuteronomy 29:18 *Make sure there is no man whose <u>heart turns away</u> from the LORD*

Deuteronomy 30:6 *The LORD will <u>circumcise your hearts so that you may love him</u> with all your heart and with all your soul, and live.*

There should be no question that Moses is saying the Heart controls, loves, trusts, promotes generosity, contains memory, is capable of understanding with reason, works in close concert with our soul, and is integral for acceptance of and interface with God. Essentially, the Heart-Brain not only does the control of physical elements needed for proper functioning of the body and its blood-flow, but it also has another very important characterization that our head brain simply cannot do.

I still hear some of you thinking he was simply being poetic like we are today; saying "I love you with all my Heart". Let me put something out there that goes against that by saying the Egyptians and many African nations thought the "liver" controlled emotional desire, while the heart provided all wisdom and communication between the Ba, Ka, and Shut. Like the Egyptians, the Jews believed the Liver was the largest and most powerful organ of the body capable of supporting love. The Egyptians were not trying to be cutsie, they truly believed this. Moses did not change all the liver feelings to heart feeling to get back at the mean Egyptians. He knew things about the heart that were not completely realized by the African nations. I know you are thinking, how can that be when he was raised by the, *"love you with my liver"* Egyptians until he was 18 years old. While the Egyptians believed the heart held the three parts of our existence together, their understanding was superficial and the liver was believed to be the compassionate organ. The greatest Roman physician name Galen described the common knowledge that both the heart and the liver were the sources of love almost 2000 years ago. This knowledge was known in all of the Middle East and Egypt. The Arabic word for liver is frequently used in Egyptian poetry and song lyrics to express love so let's look at the Bible. Instead of the Liver, it describes the bowels, but at that time that included the liver.

*I John 3:17-But whoso hath this world's good, and seeth his brother have need, and shutteth up his **bowels of compassion** from him, how dwelleth the love of God in him?*

*Colossians 3:12-Put on therefore, as the elect of God, holy and beloved, **bowels of compassion**, kindness, humbleness of mind, meekness, long-suffering;*

Just like the apostles John and Paul indicated above, the bowels were the compassionate part of our bodies, but Moses and most of the rest of the writers had an expanded awareness of the heart function. Possibly, Moses had seen ancient texts describing the truth in Heart capabilities, but I'd like to think God told him first hand. -------"*Moses! Let me tell you about my fantastic heart--------*", he might have started while Moses was viewing a burning bush or on top of a smoke-filled Mount Sanai. How ever it happened, Moses had learned what Jobab had found out over 200 years before Moses came along.

Jeremiah Writings

Jeremiah was a great historian and prophet. He had read the details established by Moses, but somehow, he gained even more insight. It is believed Jeremiah wrote the 4 books of Kings now split into Samuel and Kings as well as the book Jeremiah and Lamentations. He had similar things to say about the Heart Brain.

1 Samuel 2:1 *My [Hanna]* <u>*heart rejoices*</u> *in the LORD;*

1 Samuel 4:13 *Eli sitting on his chair by the side of the road, watching, because his* <u>*heart feared*</u>

1 Samuel 12:20 *Samuel replied,* <u>*serve the LORD with all your heart*</u>*.*

1 Samuel 16:7 *LORD said to Samuel, People look at the outward appearance, but* <u>*the LORD looks at the heart*</u>*."*

1 Samuel 17:28 *I [Eliab talking to David] know how conceited you are and* <u>*how wicked your heart is*</u>*; you came down only to watch the battle."*

2 Samuel 6:16 *Michal* <u>*despised him [David] in her heart*</u>*.*

1 Kings 2:44 *The king also said to Shimei, "*<u>*You know in your heart*</u> *all the wrong you did to my father David.*

1 Kings 3:6—*he [King David]* <u>*was righteous and upright in heart*</u>*.*

1 Kings 3:9 *give your servant* <u>*a discerning heart*</u> <u>*to govern your people and to distinguish between right and wrong*</u>*.*

28

1 Kings 11:9 *The LORD became angry with Solomon because <u>his heart had turned away</u> from the LORD.*

1 Kings 14:8 *You [Solomon] have not been like my servant David, who kept my commands and <u>followed me with all his heart</u>, doing only what was right in my eyes.*

Jeremiah 9:26 *even the whole house of Israel is <u>uncircumcised in heart</u>."*

Jeremiah 17:9 *- The <u>heart is deceitful above all things</u>, and <u>desperately wicked</u>: who can know it?*

Jeremiah 17:10 *- I the LORD search the heart, I try the reins, even to give every man according to his ways*

Jeremiah 20:9 *<u>his word is in my heart like a fire</u>,*

Jeremiah 20:12 *you [God] who examine the <u>righteous and probe the heart</u> and mind,*

Jeremiah 23:9 *My <u>heart is broken</u> within me*

Jeremiah 49:16 *The terror you inspire and <u>the pride of your heart have deceived you</u>,*

Jeremiah told us the heart could be prideful or broken. It works with the normal brain to establish righteousness, be discerning, and it can follow the wishes of God. Unfortunately, it can also turn away from God, despise those around you, and cause wickedness in a person. Additionally, we find the heart can rejoice, fear, and serve God. Not only can we see a level of consistency, we find that even after a thousand years, the ideas about how powerful the heart-brain was had not changed. Let's go on to Kings David. After all; David was extremely close to God so we should get insights.

David's View

Of course, God said David was a man after his own heart which may help us sense a heart that was made in the image of God's Heart. Again, we find that David knew secrets he was not getting from "the normal reality". David talked with God a lot and God probably let him in on how important the heart was to our normal and spiritual existence.

Psalms 34:18 - *The LORD [is] nigh unto them that are <u>of a broken heart</u>; and saveth <u>such as be of a contrite spirit</u>.*

Psalms 51:10 - *Create in me a clean heart, O God; and renew a right spirit within me*

Psalm 51:17 *My sacrifice, O God, is a <u>broken spirit; a broken and contrite heart</u> you, God, will not despise.*

Psalm 58:2 <u>*in your heart you devise injustice,*</u>

Psalm 64:6 *Surely the <u>human mind and heart are cunning</u>.*

Psalm 77:6 *My <u>heart meditated and my spirit asked</u>:*

Psalm 90:12 *we may gain a <u>heart of wisdom</u>.*

Psalm 101:4 *The <u>perverse of heart</u> shall be far from me;*

Psalm 101:5 *whoever has haughty eyes and <u>a proud heart</u>, I will not tolerate.*

David believed the heart can be happy, contrite, anxious, lustful, perverse, and broken. It can initiate injustice and be cleaned of its badness. The heart can be proud, wise, and almost completely control your actions. While these things expand what we know, they certainly go along with the ideas of Moses, Job, and Jeremiah. What did David's son think?

Solomon's View

When God asked Solomon what he wanted in all the world he told God he wanted the wisdom needed to rule the Jews and God made his wisdom greatest in the land so we might find out differences in how he viewed the heart. Would he think the Heart was a pumping machine?

Proverbs 3:5-6 - *Trust in the LORD <u>with all thine heart; and lean not unto thine own understanding</u>.*
Proverbs 4:23 <u>*Above all else, guard your heart, for everything you do flows from it.*</u>

Proverbs 6:25 <u>*Do not lust in your heart*</u> *after her beauty or let her captivate you with her eyes.*

Proverbs 12:25 <u>*Anxiety weighs down the heart*</u>,

Proverbs 14:13 *Even <u>in laughter the heart may ache</u>,*

Proverbs 15:30 *Light in a messenger's eyes brings <u>joy to the heart</u>,*

Ecclesiastes 2:10 *I refused my <u>heart no pleasure</u>. My <u>heart took delight</u> in all my labor,*

The heart can be happy, anxious, and lustful. It can be cunning or wise or it can sense great "heartbreak" and sadness. The heart can understand things beyond what our head-brains can understand. The heart must be guarded and we should meditate with it. Again, we see each of the writers are telling us about a very powerful Heart-Brain. Solomon was the wisest man every in the world so long as he stayed away from his 1000 women. As we go through the miraculous capabilities of the heart we will find it become hardened to things outside our "self". While a number of things strength the self and destroy the communication with the soul and spirit, the Three most described are self-aggrandizement, survival-fear, and sexual lust. With a thousand women around, I think you can understand that Solomon was not always wise.

Other Prophets

While we could probably go to each of the books of the Bible and pull out descriptions of the Heart-brain, we find interesting references in books written by Ezra, Isaiah, Ezekiel, and Hosea. The Prophet Ezra is believed to have written the books of Chronicles, Ezra, and Nehemiah so let's look at those first.

2 Chronicles 30:19 *who sets their heart on seeking God—*

2 Chronicles 32:25 *Hezekiah's heart was proud and he did not respond to kindness therefore the LORD's wrath was on him.*

2 Chronicles 32:26 *Hezekiah repented of the pride of his heart.*

Nehemiah 9:8 *You found his heart faithful to you.*

Like the others, Ezra told us the heart must seek God if we are to find salvation rather than our head-brain. It can be faithful or it can be proud, but it can also repent of its pride. All in all, Nehemiah thought the Heart-brain was a powerful part of our personage and we can assume the others thought the same thing.

Ezekiel 11:19 *- And I will give them one heart, and I will put a new spirit within you; and I will take the stony heart out of their flesh, and will give them a heart of flesh:*

Ezekiel 24:25 *the* <u>*delight of their eyes, their heart's desire*</u>, *and their sons and daughters as well—*

Ezekiel 25:6 <u>*rejoicing with all the malice of your heart*</u>

Ezekiel 36:26 *I will* <u>*give you a new heart and put a new spirit in you; I will remove from you your hardened heart and give you a heart of flesh*</u>.

Isaiah 6:10 <u>*Make the heart of this people is hard;*</u> *Otherwise they might* <u>*understand with their hearts*</u>, <u>*and turn and be healed.*</u>*"*

Isaiah 13:7 *every* <u>*heart will melt with fear*</u>.

Hosea 10:2 *Their* <u>*heart is deceitful*</u>,

Hosea 11:8 <u>*My heart is changed within me; all my compassion is aroused*</u>.

In their understanding, the heart can be hardened, fearful, deceitful, hateful, and <u>see things</u>. Additionally; <u>the mind of the heart can be changed by compassion</u> or by entry of God. Just imagine someone thinking the Heart <u>could see without eyes</u>. What we are finding out today is that the Heart-brain can see, but it sees a somewhat different set of E-M frequencies than our eyes. The heart can see what the eyes cannot and the eyes can see what the Heart cannot.

God's Heart-Brain Descriptions

Like I stated earlier, one could possibly attribute strange descriptions about the heart to the writers simply being misinformed, but when we read what incarnate God had to say, we find that the earlier writers were unbelievably aware of the heart having its own brain that was intimately attached to the soul. Let's review a small number of the verses attributed to the Incarnate God himself. It is true Jesus [or Joshua] as 100% man, but he was also 100% God incarnate as he had been earlier in the Old Testament when he talked to Abraham as a man and incarnate God. We can believe he gave some of these insights to Abraham directly when he was visited in Canaan and that is why so many Old Testament descriptions show a clear understanding about our marvelous Heart and Heart brain. That being said, many of the following references are repeated multiple times in the New testament to insure we understand.

Matthew 5:8 Blessed are the <u>pure in heart, for they will see God.</u>

Matthew 5:28 anyone who looks at a woman lustfully has already <u>committed adultery with her in his heart</u>.

Matthew 11:29 I am gentle and <u>humble in heart, and you will find rest for your souls</u>.

Matthew 12:34 how can <u>you who are evil say anything good? For the mouth speaks what the heart is full of.</u>

Matthew 13:15 For this <u>people's heart has become calloused; otherwise they might understand with their hearts</u> and turn, and I would heal them.'

Matthew 13:19 When anyone hears the message about the kingdom and does not understand it, <u>the evil one comes and snatches away what was sown in their heart.</u>

Matthew 15:18 <u>things that come out of a person's mouth come from the heart</u>, and these defile them.

Matthew 15:19 out of the <u>heart come evil thoughts—murder, adultery, sexual immorality, theft, false testimony, slander.</u>

Mark 12:33 To <u>love him with all your heart</u>, with all your understanding and with all your strength,

Acts 1:24 Lord, <u>you know everyone's heart.</u>

Acts 2:26 <u>my heart is glad</u> and my tongue rejoices

Acts 5:3 Peter said, "Ananias, how is it that Satan has so <u>filled your heart that you have lied to the Holy Spirit</u> and have kept for yourself some of the money

Acts 15:8 <u>God, who knows the heart,</u> showed that he accepted them by giving the Holy Spirit to them.

Acts 16:14 The Lord <u>opened her [Lydia's] heart to respond</u> to Paul's message.

So, in the Bible, God tells us the Heart responds to information, can be filled with Satan, and can be understood to know things. It can love totally, and lust totally. It can defile one's words as they are spoken. It can be calloused or understanding. It can establish humility or purity but also it can aid in evil feelings. These are certainly not the attributes of a muscle. God was not a liar nor did he simply spout poetry to make things sound good. Certainly, he established parables to help people understand things that would be difficult for them to pick-up, but the parable characterizations were always truthful physical characterization. God was trying to tell us how important the heart brain was to our existence, but how could the heart do all these things?

Paul's Writings

Paul was perhaps the greatest disciple of Jesus. He wrote more of the New Testament books than all the others and his purposed request to be sent to Rome for trial certainly shows he was willing and finally did die for Jesus, so my feeling is we should be able to use his words concerning the Heart-brain to understand it a little better. A number of the insights from Paul were outside the normal feelings of many of the early Christians, but every time someone looks for insights, Paul's discussions fill in blanks when other disciples were somewhat lacking, so let's see what Paul had to say about the Heart brain.

Romans 2:15 *The requirements of the law are written on their hearts, their consciences* [Conscience of the Hearts] *also bearing witness, and their thoughts sometimes accusing them and at other times even defending them.* This is not saying people accuse themselves, but that the heart accuses the head-brain showing how very important the Heart is. What we will find is when people harden their hearts, they no longer feel these accusations.]

Romans 2:29 *circumcision is circumcision of the heart, by the Spirit, not by the written code.* [We will have to look at circumcision later]

Romans 6:17 *you have come <u>to obey from your heart</u> the pattern of teaching that has now claimed your allegiance.* [This is substantially different than obeying with your head-brain.]

Romans 8:26 - *the Spirit helps us in our weakness. We do not know what we ought to pray for, so the Spirit intercedes for us through wordless groans. And <u>he who searches our hearts knows the mind of the Spirit,</u>* [This shows close interrelationship between the Spirit and the Heart-brain.]

Romans 9:2 *Sorrow and unceasing <u>anguish in my heart</u>*

Romans 10:8 *The <u>word</u>* [spiritual word of God] <u>*is near you; it is in your mouth and in your heart*</u>

Romans 10:9 *<u>believe in your heart</u> that God raised him from the dead, you will be saved.*

1 Corinthians 4:5 *He will expose the <u>motives of the heart</u>.*

2 Corinthians 2:4 *I wrote you out of <u>great distress and anguish of heart</u> and with many tears,*

2 Corinthians 8:16 *Thanks be to <u>God, who put into the heart of Titus the same concern</u> I have for you.*

2 Corinthians 9:7 *give what you have <u>decided in your heart</u> to give.*

Ephesians 1:18 *I pray that the <u>eyes of your heart may be enlightened.</u>*

Colossians 3:22 *Slaves, obey your earthly masters with <u>sincerity of heart</u>.*

1 Timothy 1:5 *love, which comes from <u>a pure heart</u> and a good conscience and a sincere faith.* This is a very important verse in that it is talking about "true love" of someone besides "self". This can only be achieved by empowering the soul by means of a communicating heart.

Philemon 1:20 *Refresh my heart in Christ.*

Let's see where we are as Paul describes the heart. The heart causes purity, decides direction, can be distressed or pained. It remembers words, plans, establishes belief, sees things, expresses itself, establishes obedience, and can be circumcised by the spirit. We'll look at this circumcision later.

Hebrews [by Paul]

While we don't know for certain who wrote Hebrews, it is believed it is the product of Paul, however, it may have been written by Apollos or one of Paul's helpers who were imprisoned with him for a time. Anyway, we would expect similar discussion and more insight concerning the Heart brain and we are not disappointed.

Hebrews 3:12 *None of you has a sinful, unbelieving heart that turns away from the living God.*

Hebrews 4:12 *The word of God penetrates even to dividing soul and spirit; it judges the thoughts and attitudes of the heart.*

Hebrews 10:22 *Draw near to God with a sincere heart having our hearts sprinkled to cleanse us from a guilty conscience.*

This book says it very directly; the heart can establish sincerity and can diminish a guilty conscience or it can be sinful and unbelieving. The heart thinks and stabilizes our attitudes. All this stuff sounds fanciful, crazy, mind-boggling, and scary. Could it all be true? To find out why the Bible provides 513 verses on the fantastic workings of the heart, and to see how in the world the Heart could do all of these things, we will turn to the new science of Heart Brain Biogenetics. Before we get to that, I'm thinking some you are still thinking brains belong in the head so let me introduce you to distributed brains as it is very common for brains to be distributed around the bodies of animals.

Distributed Brains

Some of you already know that really large animals have secondary brains to help speed up information routing. We also find smaller animals using a similar distributed brain function to speed up interactions or only have sensing functions where needed.

Dinosaurs-The larger herbivorous dinosaurs like diplodocus and even down to the size of a stegosaurus had a large neuron cluster somewhere along the length of their spine to help communicate signals between their primary brain and their extremities. The heart brain would be somewhat similar. The image shows the location of the butt-brain on a Stegosaurus and a Brachiosaurus.

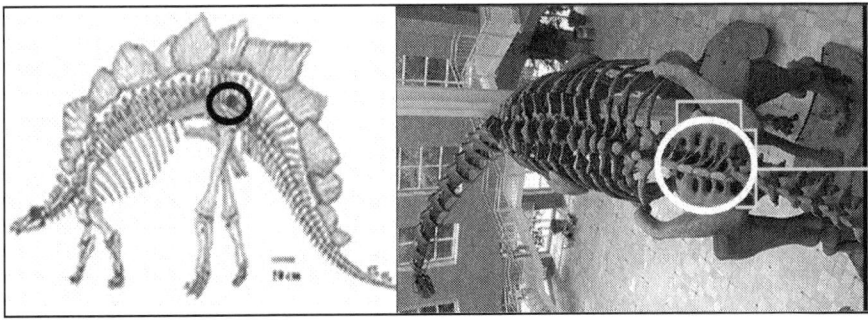

The Box Jellyfish has 5 brains, but the 4 rhopalia-tentacle brains much smaller that the main brain with only 1000 or so neurons in each tentacle.

An octopus has 9 brains with an additional brain in each of its 8 tentacles and there is very little connection between each brain so that an octopus must look to be sure where the other brains placed the tentacles.

Sea Urchins have no centralized brain and rely on a distributed sensing system that allows them to move, find food, and whatever else a sea cucumber wants to do.

Sea anemone- has a tiny brain in each of its massive number of tentacles. The main body doesn't need much instruction as its anus and mouth are the same opening so the multi-tentacle brains do most of the thinking.

Sponges have no centralized brains, but they still have sensing elements on its skin to allow it to find and eat food. Here is an interesting thing they do with a secondary brain. If they sense some irritant, they can sneeze for up to a hour trying to remove whatever got inside them. OK it is a stupid animal for sure.

Starfish like an octopus have sensors in each of its 5 appendages that sense food and the world around it, but it has no centralized brain. It is a 5 foot-brained animal.

Sea cucumbers have no centralized brain, but ejects a sticky substance out its anus if attacked or sometimes it simply throws up substantial portions of its guts hoping the attacker will be startled.

Ants- To improve and speed up sensing, each of an ant's antennae has its own set of brain-Neurons. The antennae can work separately from the ant to sense out issues with the surrounding area. Something that is interesting about the Ant is that its brain is similar in size to our Heart brain so let's see what that brain does.

Ant Brains

Let's look at the ant. The reason is that the <u>ant brain and the heart brain are roughly equivalent in physical size</u> but the Heart-brain is infinitely more important to us in ways that even extend outside this universe. Like the brain of the heart, the Ant has brain stuff distributed in its head and antennae. With its tiny brain, it does the following: It learns to eat, walk, climb, jump, sting, cut down pieces of leaves, carry massive things 4 times as long as them, and 3 times as heavy. This means it must continuously compensate its weight and balance as the wind and very high center of balance tries to flip it over. Leaf cutters don't cut grass and leaves for food. Instead, they bring the grass into enormous colonial farms where the grass is used to farm. Masters at fighting these ants learn how to take down and kill just about any intruder and they use well establish teamwork to destroy enemies many times their size. All this is done with the tiny brain shown below.

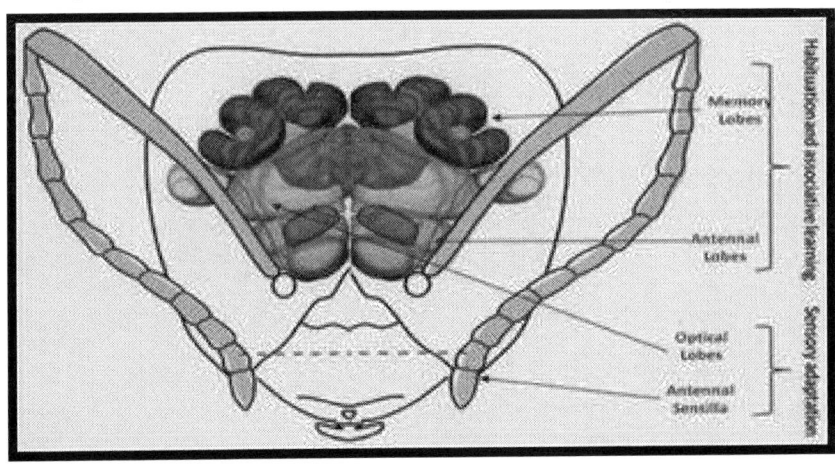

Special capabilities- The ant brain allows the manufacture and detection of caustic liquids for protection, detection, and finding the way to and from potential harvests.

Trail blazers- continuously emitting the liquids to build roadways to the potential leaves to be harvested, their tiny brain works all that roadway building. And the harvesters "sense" the trail using their versatile brain so that they can find their way to home and back again.

The ant architects design and establish construction of colonies that often extend 100 feet deep. Given their size this is the equivalent of a 3600-story set of buildings in a single colony that would be 17 miles in diameter with housing for 8 million occupants and a massive farm. The internal tunneled roadways are all secured with cement mixed in place and deposited as the work continues.

Hearing-The ant also hears with its long antennae as information about vibrations is channeled to the brain for interpretation---even without ears. The ant uses the antennae as its heat, and touch sensors as well as the brain to feel the environment as well.

Memory- The ant knows his tribe and particular duties, He remembers where he needs to go and where he needs to carry his load and even the art of sex when necessary.

Community pride- The ant will do anything needed to protect the colony. In times of severe flooding massive colonies have been known to build an ant body raft that allows some of the colony to survive and rebuild.

The Horticulturalists of the society- design and maintain massive farms to continually feed the 8 million entity colony and establish ways to keep the fungus being farmed alive and juicy. Each leaf brought in to the colony is chemically tested by the ants to insure it will make fungus thrive and those that are problematic are quickly discharged.

Waste Managers- This group continuously picks up parasitoids and other waste and remove those components form the colony daily as their brain sorts out what should or should not stay.

As phorid flies continuously try to lay eggs in leaf cutter ant heads, small ants go with the collectors to stay on the heads and ward off these intruders. Their brain tells them exactly what to do if intruders try to insert eggs.

The well-established civilization includes many specific castes with various assignments of each depending on social need These include the military, homemakers, architects, farmers, and cutters. The collage below shows from left to right close colonization, carrying a huge leaf, finding the trail home, the massive size of a typical colony and the leaf farm. Those are real people working to excavate the well-designed "city".

While some of these things are of no interest to the Heart-brain, our Heart-brain does similar, complex things, and much more, but I wanted you to understand that even a very tiny brain has enormous functionality. To understand something about our Heart-brain we must first start with a relatively new science called Heart-Brain Biogenetics. This is the seed of our understanding the Heart-brain and understanding some of the more unusual parts of our Bible.

Heart Brain Biogenetics

I know all those descriptions sounded religious and spiritual and all that, but what we are finding out about the heart and its brain are just about as "religious" as that presented in the Biblical testimony, but now it is being found and tested in a laboratory. The heart seems to be able to do all of this stuff. Scientists have now found that the heart is more than just a pretty face. It has its own 'stand alone" brain, endocrine system, communication/sight, and control over our bodies and sometimes, even our head-brain. Instead of being a massive clump, like the head-brain, the neurons of the heart-brain are localized throughout the main vessels of the pump part and the various sections are all interconnected.

- 20% of these neurons have to do with mechanical information (pressure etc),
- The other 80% is sensitive to chemical substances (hormones, neurotransmitters) and sets up actions that can allow the Heart to see.

I know that sounds like we ran out of brain, but we will see the Heart-brain is way more mysterious. This information gathers in the little brain in the heart, where it is integrated and used for local decision making. The Intrinsic neurosystem of the heart is also directly connected with skin, lungs and other organs so the Heart can control these body parts much faster than the head-brain.

While the Heart has a brain, it is a tiny thing by any definition so having all the previously described capabilities seems improbable when compared to the head –brain that is massive. The Heart brain elements are made up of about 40 thousand neurons while the head-brain is made up of something like 18 billion. While we are told no one uses more than 10% of our brain, we actually use much, much less than that and substantial amounts of our head-brain are bogged down in interpreting E-M signals that are turned into chemical charges by the sight sensor eyes. The brain takes the mesh of varied E-M signals and creates vision. The head-brain also takes vibrating pressure signals that are turned into chemical charges and builds appropriate meaningful "sounds". The head-brain also establishes characteristic muscle contractions needed to support vibrating our vocal cords for voice modulation and even "singing" for some of us.

Head-Brain Freedom

Generally speaking, the Heart-brain doesn't need to do any of that. If it needs that information, the Head-brain can get it like going to a library or the internet or using a calculator rather than trying to do math, so maybe, 40 thousand neurons not required to do the daily bookkeeping chores is enough to accomplish some remarkable effects. While we might thing of the Head brain as this master memory of carnal events, saying, songs, etc, the Heart-brain could be thought of as the glue that holds the three portions of your entity together. Separate the Soul, Spirit, and Self are not alive, so in a way, the Heart-brain allows us to be alive.

Neuroscientists today tell us the heart and its brain are far more complex than we'd ever imagined. Instead of simply pumping blood, it may actually direct and align many systems in the body so that they can function in harmony with one another.

To make this seem even more surreal, this new scientific evidence shows that the heart uses various methods <u>to send our brain extensive emotional and intuitive signals and it does the same to heart-brains of other people nearby and it may even affect the reality of the world itself.</u> The heart is in constant communication with the brain and scientists are now beginning to believe that our hearts are the "intelligent force" behind the intuitive thoughts and feelings we all experience. To understand the heart-brain we will have to learn a little more about the three entities that make us up as a person. [The spirit, soul, and self] Before we get into all of that, let's look at the more normal, heart-brain communication with our head-brain.

Heart and Head-Brain Communication

The previous definition tried to stay esoteric, but this one really hits home. *[John and Beatrice Lacey,1978]- However, following several years of research, it was observed that, the heart communicates with the brain in ways that significantly affect how we perceive and react to the world. It was found that, the heart <u>seemed to have its own peculiar logic that frequently diverged from the direction of the autonomic nervous system</u>. The heart appeared to be sending meaningful messages to the brain that it not only understood, but also obeyed.*

It wasn't until 1994 that another neurophysiologist, Dr. Armour introduced the concept of functional 'heart brain'. His work revealed that the heart has a complex intrinsic nervous system that is sufficiently sophisticated to qualify as a 'little brain' in its own right.

His report indicated the following: *The heart's brain is an intricate network of several types of neurons, neurotransmitters, proteins and support cells similar to those found in the brain proper.*

> *Its elaborate circuitry enables it to act independently of the cranial brain – to learn, remember, and even feel and sense.*

The heart's nervous system contains around 40,000 neurons. Information from the heart; including feeling sensations, is sent to the brain through several pathways. These main nerve pathways enter the brain at the area of the medulla, and cascade up into the higher centers of the brain, where they may influence perception, decision making and other cognitive processes. Like our head brain they are masses of "brain gue" called Ganglia. The images following are of these ganglia and their connective synaptic nerve fibers.

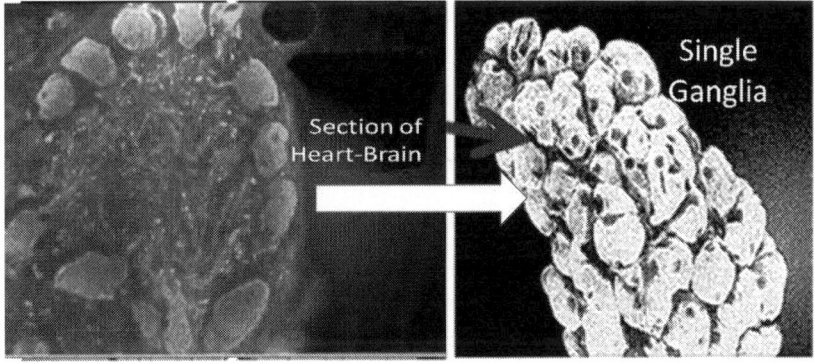

Here is a strange and possibly important part; these ganglia are distributed throughout the heart as shown again following this paragraph. This may give them the effect of providing faster, freer access, computation that the head brain. Each dot in the following diagram is another ganglia cluster. As shown there are actually 7 different brain areas in the heart. Some of the brains just works the heart muscles and the pumping rhythm or speed things up when stress, but there is a lot more to heart-brains than that.

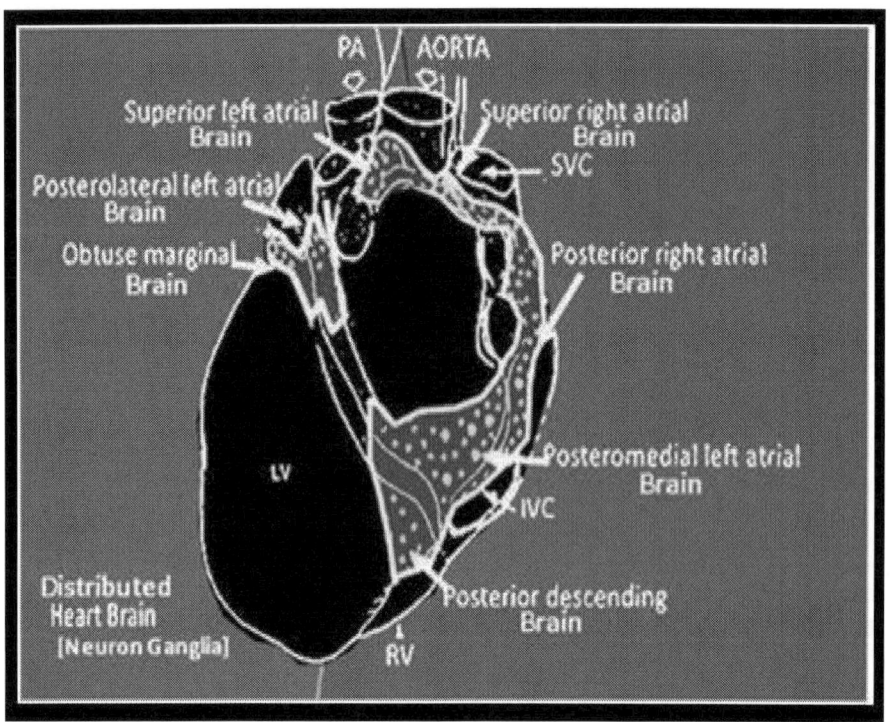

You are being so stubborn, let me give you some more verses. I know all this seems radical as you have not been taught anything about the Heart-brain in classrooms and most shy away from the possibilities as the head-brain has many times as many neurons as the Heart-brain. Surely the Head-brain would be the main processor of information in the real world as it is directly tied to all the major sensors. Ears, Eyes, touch nerves, smell sensors, heat/cold sensors, and pain sensors and it is tied directly to the response elements of most of the body components- muscles, skin, and most of our guts.

Gut-brain

Speaking of Guts, we need to consider the gut-brain or what is commonly called the Stomach-brain. For one thing it is way bigger than the Heart brain. Consisting of 500 million neurons, in size it is a thousand time larger than the heart brain and it is distributed along the entire digestive track. Like the Heart-brain, it can operate autonomously, but there really is no major sensors except hunger pains and we have all had that butterflies in the stomach feeling as this third brain of the body works to make itself known. While controlling some levels of mood and anxiety, the Heart-brain is much more functional and some believe the heart uses the gut-brain for temporary memory storage from time to time; besides, guts are not all that are needed to react and interface properly with our body, we need hormones.

Body Hormone Controller

Besides having its own brain, it has a wide assortment of ways to affect the body. Certainly, it can change the heart-rate, constrict blood vessels and change the rest of the body. As it commands a hormonal Army.

Before this time, the heart was considered a pump, but the heart was reclassified as an <u>endocrine gland</u> in 1983 because it controls a wide number of hormones.

Atrial Natriuretic Factor Hormone-This is the <u>main hormone produced and released by the heart</u>- "atrial natriuretic factor" (ANF) We didn't even know about it until 1983. This hormone exerts its effect on the blood vessels, on the kidneys, the adrenal glands, and on a <u>large number of regulatory regions in the brain</u>. The heart allows us to get the blood we need not only by simply pumping it but also controlling how large our blood vessels expand.

Noradrenaline and Dopamine- The impressive heart was also found to contain a cell type known as 'intrinsic cardiac adrenergic'' (ICA) cells. These cells synthesize and release noradrenaline and dopamine that establish state of mind in our bodies.

Oxytocin-More recently, it was discovered that the heart also secretes oxytocin, commonly referred to as <u>the 'love'</u>

or bonding hormone. In addition to its functions in childbirth and lactation, recent evidence indicates that this hormone is also involved in cognition, tolerance, adaptation, complex sexual and maternal behaviors, learning social cues and the establishment of enduring pair bonds. Concentrations of oxytocin in the heart were found to be as high as those found in the brain from studies in 1986. This means ½ of all the feelings of love of your body come from the heart. Isn't that weird?

Instead of saying I love you with all my heart, you should be saying I love you BECAUSE of my heart just like the Biblical writers told us.

Other Neuro-Transmitters- Besides direct hormone interaction, it has been found that the heart contains cells that synthesize and release norepinephrine, epinephrine and dopamine, which are neurotransmitters once thought to be produced only by neurons in the brain and ganglia.

Hopefully you are starting to see a picture of a master controller, but the next section will help.

E-M Messaging to the Brain

Besides sending messages along the normal nerve pathways and adjustment of hormone levels, our heart has another tool. It sends radio-waves or E-M waves and with those signals, the heart sends messages. The heart actually syncs up with the head brain or gets it to sync up with the heart with what is called "Coherent heart rhythm patterns". These are sent to the brain to allow the brain to calm down and think and researchers have seen this effect happen in the lab. The effect is often experienced as <u>heightened mental clarity, improved decision making and increased creativity</u>. Additionally, coherent input from the heart tends to facilitate the experience of <u>positive feeling states</u>. This may explain why <u>most people associate love and other positive feelings</u> with the heart and why <u>many people actually feel or sense these emotions in the area of the heart</u>. The heart rate is never constant. Instead it changes from beat to beat, either in a rhythmic patterned or in a chaotic pattern. An example is shown next.

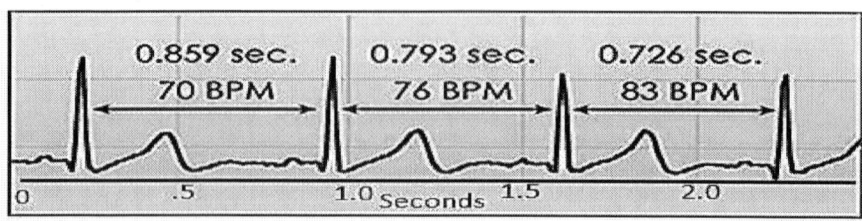

If you just look at the rate variations you start seeing something interesting. The following diagram shows heart rate fluctuations when a subject was highly frustrated as tested at the HeartMath Institute who are one of the premier heart capability and heart modification labs in the country.

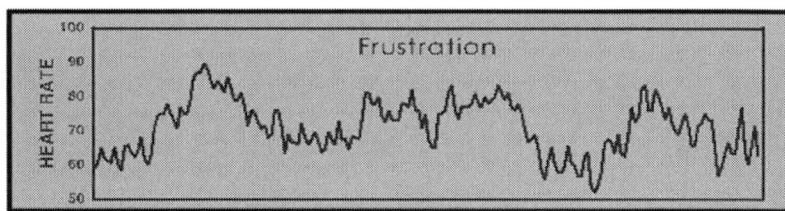

This is not much different than what we might call normal as most people are generally experienceing a level of frustrational all the time this "normal" is shown next. Please notice that respiration, and weart rate variability similar to that shown above are somewhat chaotic. The first graph, in the following collage, simply shows that respiration is not synced to these 2 elements of blood flow communication. The second graph is the same as I showed before [HRP] showing heart rate variability [HRV] over time. The last chart simply looks at the variability of blood pressure over time which also establishes its own rhythm. The last graph shows how long each of the changes in blood pressure last over time. While it seems to be similar to HRV, it is not the same as high pressure can certainly be found with low pumping rates. What these show is that if something could interpret all of these changes in blood pressure and heart rate changeability, one could harness a level of communication. The Heart is trying to talk. All it needed was a transmitter and I have described that previously as our massive collection of blood vessels building magnetic fields each time the heart rate changes and each time the blood pressure changes. To decode these messages, we simply need another heart brain nearby.

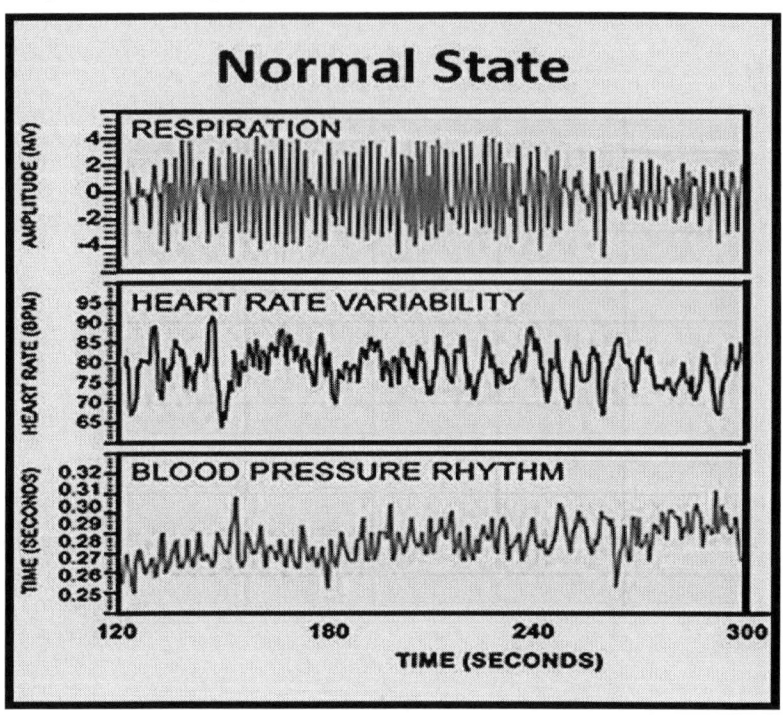

What do Messages look like?

Hormonal messaging affecting the other organs and the Brain look like increases and decreases of hormones to signal events and conditions or to modify current conditions of various organs and the heart. These modifications would be very slow and generally the effects take time as well. While we know these are important, what I'm talking about here are the electro-Magnetic messages [low frequency radio waves] produced by intricate modifications in blood pressure and blood streaming throughout the body. In a rest state, the main frequency can be established as the "heart rate" as shown again below this pulsing occurs at about 1.3 Hertz rate that is modulated from about 0.8 to 2 Hertz.

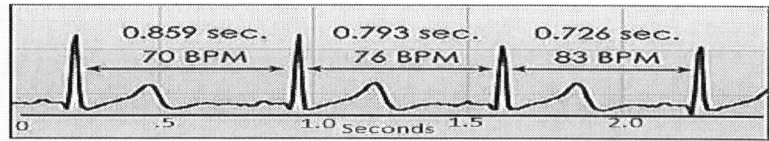

Modulated "Coherence" Pattern

While this modulated heart-rate establishes one part of the messaging, the variability of the pulse rate established another type message. If the rate varies at some regular pattern, we can call this coherence. The heart attempts to establish this "resonance" with the brain all the time to allow it to think more clearly and reduce its stress. The following graphs shows coherence as a rate going between about 60 and 50 beats per minute [BPM] at about ¼ Hertz rate but the heart doesn't stay this way. After the Heart does its magic during calming sessions, the heart rate typically begins changing in a rythmic pattern similar to that s shown below. In this state, the brain is sharper, fast and the feeling of love abounds. Reaseachers call this phenomenon Coherency.

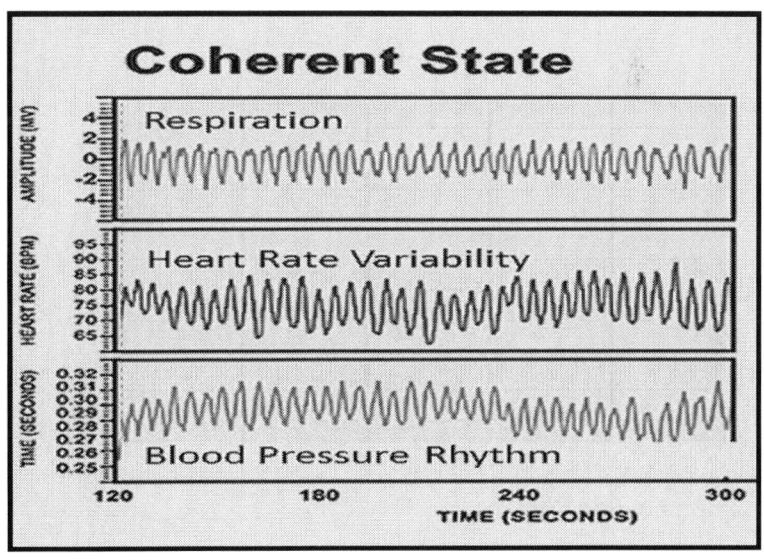

From HeartMath Institute Clinical studies and those monitored by the institute we find the following benefits of this Coherence and incidents of same.

- A 20% reduction in blood pressure Blood Pressure found in Hypertensive Employees
- A 30% increase in Type 2 Diabetes quality of life metrics;
- A 14% improvement in functional capacity from Congestive Heart Failure patients and 75% of patients had significantly fewer Heart Arrhythmias and 20% were able to stop medication altogether.
- A 50% decrease in airway impedance of Asthma patients

Depending on internal and external events, it changes wildly and the more erratic the changes in blood pressure are, the more stressed one appears to be.

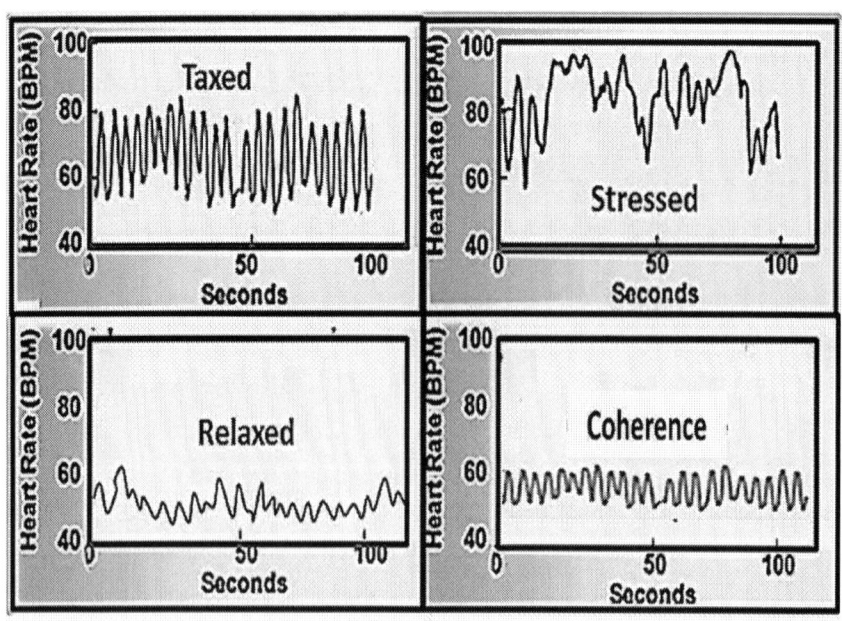

The HeartMath Institute has tried to standardize the heart messaging language so let' use their descriptions here. Anything that does not look like the regular changes of coherence is called----wait for it--- non-coherence. As a note, simply being relaxed is not having the heart and brain in coherence, it just means the general blood pressures are typically lower. This whole testing of the variability of the blood pressure is called heart rate variability (HRV) and our HRV is usually greatest when we are young. While having a low HRV sounds like our head-brain would work the best, if it is too low it sometimes shows some abnormality that may indicate a premature death or health issues. When we put the HRV on a times graph it can be called the Heart Rhythm pattern [HRP]. I know this is all confusion, but these abbreviations help us understand Heart-Brain communication.

The next graph is fairly special to HeartMath as it is a Fourier transform of the HRV. A Fourier transform simply represents something that happens over time as how often it changes over time [This is called the Frequency domain. It's kind of like the difference between FM radio where data is carried by changing frequency instead of AM radio where the data is carried by the amplitude of the signal directly. It seems that Heart brain transmissions might be sensed as both so we should look at both.

The graph is of Power Spectral Density (PSD) of the variations of blood flow and here is why this was done. Notice that normally, the heart messaging has a fairly low "Power" level, but whenever the heart gets into a coherence mode. This can be with the head-brain or with an adjacent heart-brain of a person. If coherence is reached, the power available in the EM communication [low frequency radio waves] has 7 times the power in the example following.

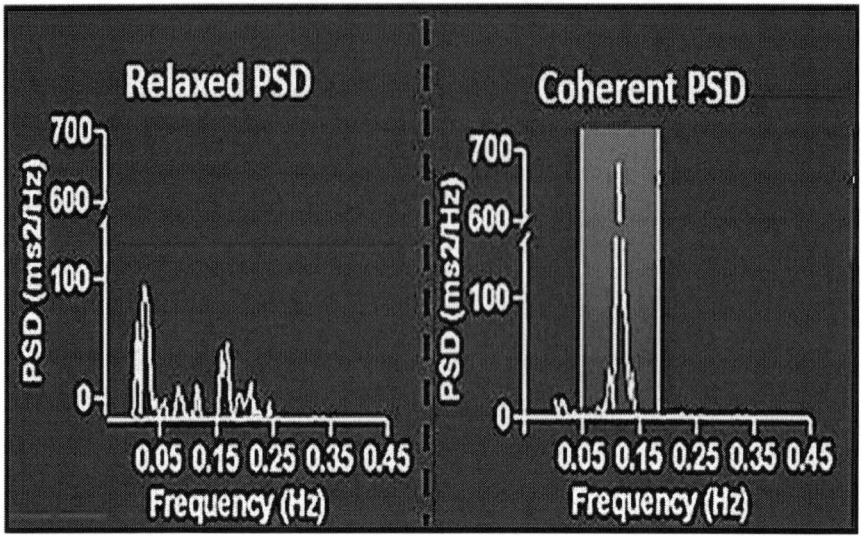

Essentially, a heart can talk to another heart at a much longer distance if they are coherence. While these might be loved ones, it is noticed that a baby and its mom can get into this mode of communication and even a horse and its rider seem to link up this way if they like each other. In a same kind of experiment, we find the following:

- Two people stand five feet apart, and subject two's brainwaves synchronize with the heartbeat of the other.
- The hearth rhythms of dog and boy synchronize when love is sent, <u>while they are in separate rooms</u>.

Unlike relaxation, coherence does not necessarily involve a reduction in HRV, and may at times even produce an increase in HRV relative to a baseline state. Here is the interesting part for me. What was found was that the blood pressure ebb and flow variation when plotted was a sine wave between 1 and 1/10 of a hertz. This subsonic region is of interest in many areas of non-heart related brain manipulation. The main ones of interest here are "infratonics" and something called "transcranial magnetic stimulation". Both use ultra-low frequency vibrations to calm the brain, eliminate pain, place the brain into a heightened sense of reality, and even cure some medical conditions. While these seem like magic when instituted by external mechanical means, the heart evidently does this all the time. Let's look at some of these external heart messaging devices.

Subsonic Brain Manipulation

Our brains typically stay in states called Alpha, beta and Gamma which mean the main Vibrational characteristic of a brain [not in coherence] is between 8 and 100 hertz as shown in the following chart. When the brain senses coherence with eh Heart-brain, it slows down substantially. What is interesting is that if one can slow down the rhythmic nature of the heart, all types of things begin to happen. Please notice the following:

Epsilon Thinking-If the brain is operating below about ½ Hertz people sense out-of-body experiences, spiritual insight, high levels of inspiration, and a high state of meditations.

Delta Thinking- [Less than 4 Hertz] produces a hypnotic state and also has an increase in intuition as well as some healing from trauma.

Theta Thinking [less than 7 hertz] increases memory function, increases creativity, gives a profound inner peace with emotional healing, and allows for psychic imagery.

While the heart does this simply by sending the Coherence massages, researchers are now simulating the heart with external devices. Because they have to get though bone, the signals must be stronger, but the idea is to force the brain into a lower state and bones will be repaired, stress will be eliminated, and on and on. Here is the interesting part. Test labs across the country are affecting our bodies simply by changing our brain base frequencies and watching in amazement sort of like when you watch a "faith healer". It seems completely bogus, but there you have a cured person. Just image how amazed the people were when one of the disciples would simply touch them and they could see, or stand, or whatever, and they did it without a machine.

Type	Freq. (Hz)	Normal Reactions
Epsilon	<0.5	Extraordinary states of consciousness, High states of meditation, Ecstatic states of consciousness, High-level inspiration states, Spiritual insight, Out-of-body experiences, Suspended animation.
Delta	0.2 to 4 Hz	Confusion, boosting intuition, Deep sleep, Lucid dreaming, Increased immune functions, Hypnosis, Anti-aging, Increased intuition, Inner being & personal growth, Trauma recovery, Near death experience, Blissful "being" state
Theta	4 – 7 Hz	Arousal, Deep relaxation, Increased memory, Creativity, Hypnagogic state, Access to subconscious images, Reduced blood pressure, Profound inner peace, emotional healing, Inner wisdom, Faith, psychic abilities, Twilight sleep learning, Vivid mental imagery, Military remote viewing
Alpha	8 – 12 Hz	Relaxation, Meditation, Light relaxation, Positive thinking, Creative problem solving, Mood elevation, Stress reduction, Intuitive insights, Daydreams, Calm, relaxed, Lucid mental states, Tranquility, Detachment
Beta	12 – 30 Hz	Alertness, Anxious thinking, Active concentration Analytical problem solving, Judgment, Decision making, Increased mental ability, Focus, Good for absorbing information passively, Treating Hyperactivity, Sensorimotor Rhythm, Outer awareness, Arousal, Dendrite growth,
Gamma	30 – 100 +	Motor functions heightened, Boosted memory, Enhanced perception of reality, Binding of all senses, Increased compassion, High-level information processing, Natural antidepressant, Positive thoughts, Higher energy levels, Decision making in a fear situation, Muscle tension, Release of growth hormone, muscles, Recovery from injuries, Rejuvenation effects

As the effective frequencies of manipulation are subsonic in nature, researchers calling one of the devices for simulation Infratonic so no one gets confused with Subsonic, I suppose.

Infratonic Stimulation

It seems one can affect the brain either with E-M fields, like the Heart of by manually moving the brain with subsonic sound-waves. No one really knows why the brain is modified, but researchers have devised ways to force these had to manufacture sounds. To test and determine what vibrations affected the brain in what ways, there have been 2 major methods deployed currently. Both of these are "audio entrapment" methods. That is, they allow the brain to interpret much lower frequencies than those actually transmitted.

Two Testing Methods

The first method "binaural interpretation" is accomplished by sending slightly different frequencies in each ear. The brain picks up on this difference and uses the difference in its normal excitation. Let me give you an example a 100-hertz tone is put into one ear while a 100.5 Hertz tone is put in the second ear. The brain gets both signals simultaneously so it senses a difference of ½ Hertz which has been shown to change the brain patterns and cause some of the affects addressed earlier. I must admit I tried this method and I will tell you a felt like I could very well have left my body if I continued, but I got worried and stopped after a time. By the way, this difference signaling is put to music so you can't even tell the difference in frequencies in your conscious mind.

The second method is called "modulation". Similar to the other, this method simply modulates any music or tone very slightly. The brain senses the modulation and interprets that as its control function directly, without being combined in the ears.

Other methods-Certainly there have been blinking lights, mechanical vibration, and magnetic modulation methods, but the audio methods are extremely easy to accomplish. The Infratonic Qui Gong Machine, for instance, was developed out of scientific research in Beijing China which studied natural healers and found that <u>most powerful healers were able to emit a strong infrasonic (low frequency sound) signal from their hands</u>. The sound emitted from average individuals was only a hundredth as strong. The "Infratonic machine", is now used by 1% of all doctors in the United States and it is believed that it also is an audio modulating device.

Sphincter Resonance-Not everyone has gained success as can be illustrated with something called Sphincter Resonance. In the 1960s, somebody discovered the resonating frequency of the sphincter. Presumably, this team created a device later called an "Anal Sphincter Resonator". It was, supposedly, kind of like a musical organ. The idea was to intensify the "suspense" in movies whenever "Danger" was about to be portrayed. BACKFIRE and more BACKFIRE. Apparently, it caused the entire audience to soil themselves. The specific group of tones generated by this contraption has been referred to as a 'Brown Note' for some reason that I am not going into at this time. The specific notes have been lost over time, so I'm sure one of these mishaps will occur again in the future.

I would assume the Heart brain could establish this brown note if it wanted, but luckily, we do not have hearts without heart.

Wait a minute! Have you ever gotten an upset stomach and had no idea why? I told you the Heart-brain was independent which allows it to communicate with our soul and spirit, so what if it wanted to play a joke-----never mind. The images below are of a few of these artificial heart signal devices being used to reduce chronic pain and other healings.

  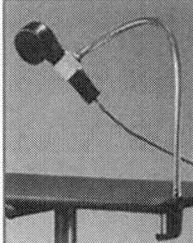

Transcranial Magnetic Stimulation (TMS)

A Dr. Keely designed something called the Krell Helmet back in the 1960s that relied on electromagnetic fields generated in the helmet to cause similar effects as the Infratonic devices with magnetics. I don't know how successful this machine was, but it would be the beginnings of something called Transcranial Magnetic Stimulation [TMS]. Rather than having a massive transmitter, this uses a massive pulsing magnet that is passed across the brain at a very slow rate which modulated an E-M field at subsonic rates similar to the Heart Brain. Today you can find these things everywhere, John Hopkins, the Mayo Clinic, and on and on we can go. Today rather than having to physically pass a magnet across the brain, machines move the fields around to do all sorts of things for the brain. Some indicate it's like degaussing the brain, and it operates like the old degaussing coils in the old CRT television sets, and possibly that is what the Heart messaging does.

TMS is defined as a non-invasive method of brain stimulation that relies on electromagnetic induction using an insulated coil placed over the scalp. Sometimes it is focused on an area of the brain thought to play a role in mood regulation and other times it is moves around larger areas. The coil generates brief magnetic pulses at an extremely slow rate. The pulse rates are kept secret, but we can imagine they are less than a couple of Hertz. The pulses are of extreme levels similar to an MRI machine, just to get though the scalp and into the brain, The Heart doesn't have to go through a skull so its job is easier. Here is the thing, these external Heart message simulators are producing fantastic results with lasting reductions in pain, and emotional stress without the side effects of drugs. The devices shown below are similar of those used at doctor's offices. Notice the last one is a handheld manual one. There are many, many types and sizes of these things being used today and most don't even know they are simulating the Heart-brain.

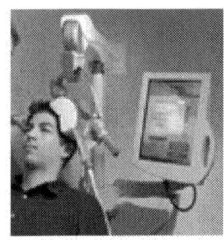  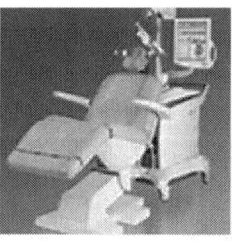 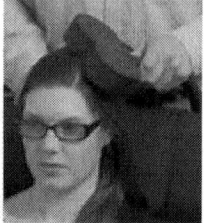

Interestingly this was first used on Rats and it was found that they not only healed wounds faster, but that they were able to learn how to trace mazes faster. The rats got smarter. I would say the Rats got wiser, but then I would sound like one of those Biblical writers describing the Heart.

Heart-Brain E-M Emitter

As I said the E-M field of the Heart brain has been determined to be 5000 times as strong as that of the head brain and it is a very low frequency. Here is the thing to know if you want to successfully emit low frequency signals; you MUST have a large emitter. One of the reasons for the Heart-brain's massive capability is the emitter itself. The Heart-brain used variations in blood flow using the entire body's blood vessel system to focus messaging as needed.

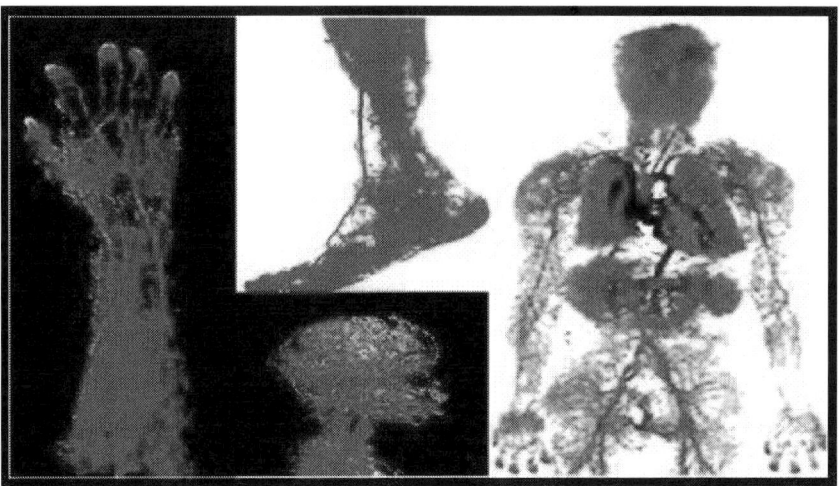

Each time there is a push of charges material it changes the magnetic field produced and when the pressure is released, another opposite phase E-M signal is produced and this emission is along the track of blood. While the head looks like a big pool of pulsating blood, the rest of the body is just about as dense as the whole body emits whatever the blood does. The images do not show the delicate capillary blood flow that just about fills in the rest of the images and we need to understand something about E-M emission here. They work off surface area so even the fine capillaries add to and generated field. Please notice something important. If you wanted to send messages to another heart-brain, one reasonable way would be to place hands on the other person as the fingertips are loaded with the high message content blood vessels. The Bible indicated that if we wanted to enhance healing possibilities, we should place our hands-on those needing healing or encouragement. This limited healing range is identified over and over again in the Bible as it continues to expand our awareness of our Heart-brain. I want you to notice something. I made a quick sketch showing a radar transmitter, TV receiver, and a faith healer placing his hands on someone. Do you see any similarities? If not, I'm sorry about my artwork. All the pulsing blood vessels pulse in unison and the signal is amplified by the sheer size of the emitter. Let's see what the Bible says again. By the way, the reason I am using the Biblical statements so much in this book is not for religion so much but because the Bible gives us a lot of information that is continuously being confirmed.

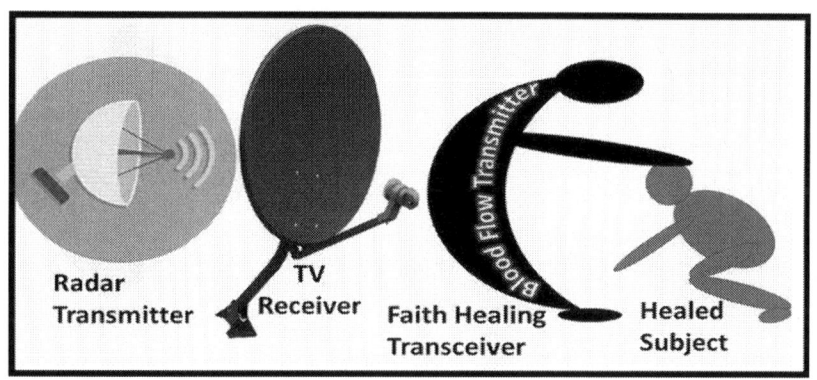

Listening to your soul or heart can help us live with more love affection and more in tune with our reality. Later we will look at problems heart transplant recipients had with respect to their new feelings some were meat lovers, now hating meat; others were homosexual and now turned straight. When you get someone else's heart, there are memories that come with the new heart brain.

Old Testament E-M Communication

Numbers 8:10 *-and present the Levites before the LORD; and the sons of Israel shall lay their hands on the Levites.*

Numbers 27:18 *-So the LORD said to Moses, "Take Joshua the son of Nun, a man in whom is the Spirit, and lay your hand on him;*

Deuteronomy 34:9 *-Now Joshua the son of Nun was filled with the spirit of wisdom, for Moses had laid his hands on him; and the sons of Israel listened to him and did as the LORD had commanded Moses.*

Messages transferred seemed to be one to help expand communication with the spirit in all three cases, but we find this communication extension had another important use in the New Testament.

New Testament E-M Communication

Matthew 19:15 -*After laying His hands on them [children], He departed from there.* [Help communication with Spirit]

Mark 10:16 -*And He took them in His arms and began blessing them, laying His hands on them.* [Help communication with Spirit]

Mark 6:5 -*And He could do no miracle there except that He laid His hands on a few sick people and healed them.* [Heal the sick better]

Mark 7:32 -*They brought to Him one who was deaf and spoke with difficulty, and they implored Him to lay His hand on him.* [Heal the sick better]

Mark 10:16 -*And He took them in His arms and began blessing them, laying His hands on them.* [Help communication with the spirit]

Mark 16:18 -*they will pick up serpents, and if they drink any deadly poison, it will not hurt them; they will lay hands on the sick, and they will recover."* [Heal the sick better]

Luke 4:40 -*While the sun was setting, all those who had any who were sick with various diseases brought them to Him; and laying His hands on each one of them, He was healing them.* [[Heal the sick better]

Acts 6:6 -*And these they brought before the apostles; and after praying, they laid their hands on them.* [Help communication with Spirit]

Acts 8:17 -*Then they began laying their hands on them, and they were receiving the Holy Spirit.* [Help communication with Spirit]

Acts 13:3 -*Then, when they had fasted and prayed and <u>laid their hands on them, they sent them away</u>.* [Help communication with Spirit]

Acts 28:8 --*father of Publius was afflicted with recurrent fever and dysentery; and Paul -<u> laid his hands on him and healed him</u>.* [Heal the sick better]

1 Timothy 4:14 -<u>*Do not neglect the spiritual gift within you, which was bestowed on you through prophetic utterance with the laying on of hands by the presbytery.*</u> *[Increase capability of seeing visions]*

1 Timothy 5:22 -<u>*Do not lay hands upon anyone too hastily and thereby share responsibility for the sins of others*</u>*; keep yourself free from sin.* [Two-way communication]

2 Timothy 1:6 -*For this reason I remind you to <u>kindle afresh the gift of God which is in you through the laying on of my hands.</u>* [Increase capabilities of people]

Forget I told you this, but after death, the feelings, emotions, love, and various memories may last in the portion of this brain that is connected with the now free soul as it might be awake to be transferred into a new heart of a new self. Before the soul finds solitude in sleep or reincarnation, it is believed one can communicate or sense characterizations assumed by a departed loved one, a loved one in a coma or even regular live person nearby. Later we will explore how these communications occur, but by many accounts, to accomplish this communication one must reduce feelings of selfish desire, prideful attitude, survival fear, and sexual lust which control our "carnal" images centered on our head-brains. Then one can begin to focus on unselfish love, absolute truth, and awareness of true reality. These are elements of our soul existence. As we focus the Heart-brain communication more on the soul we may lose a little coherence, but it is well worth it as healing the sick can be accomplished by using our heart-brains correctly.. Just like modern *"Faith Healers"* the disciples cured the sick, but many times, to enhance the E-M communication of the Heart-brain, one would place the "emitters" on the subject of communication.

Healing with Hands

The Biblical stories tell us that if we have "faith" we can heal by simply placing our hands on an injured person. But there might be more to this and the various faith healers that try to use hypnotheropy or grand utterances from a pulpit, or smashing someone on the head while in a state of trance, or building up one's CHI and sending something through a person's body to cure. All these "techniques" seem to have some success in treating a wide assortment of issues but sometimesa the "healing" is limited just like normal doctor's results using fancy herbs made in laboratories that had apparently helped a rat feel better. Let's see what these methods might really be doing.

Christian Laying on of Hands

Christian Laying on of hands is an extremely ancient method to cure disease of all types even up to death itself. The problem is it takes something called faith and the definition of that is somewhat obscure. Some thought it takes trust in God, but as I mentioned earlier, the faith described for faith healing is not the same as the Trust in God Faith needed for salvation. Therefore, many non-believers [and charlatan ministers] are still able to lay hands on people and cure them. The images below are of Catholic priests practicing it and a painting showing Jesus helping the sick.

This faith stuff seems to work very well in this application so we had better learn about it. Laying hands on the sick was a common practice in the Early Church. Throughout the New testament we read over and over again about Jesus, his disciples and charlatan ministers laying their hands on patients and curing them. In Mark 16:18, Jesus said concerning His disciples, *"they will place their hands on sick people, and they will get well"*

This form of faith is simply an understanding and interconnection between self and the perceived reality we live in—sort of like the self-actualization of Thomas Maslow, but one step more intense. By "KNOWING" one can emit this "energy" and focusing on that event one can do it and people can be cured. This is similar to Hypnotherapy in the general sense but the subject is placed in a very relaxed state before the healing is done.

Hypnotherapy

In hypnotherapy, it is suggested that in a hypnotic state, all the cells of the body are more responsive to suggestions of healing. Our brains go into a beta or slower vibration state so that it can receive Heat-brain messaging more easily.

We can see from the hypnotherapy session shown next that the hands are placed very close or on the body at different times, even if the "patient" is not looking at them. This has nothing to do with hypnotism, it's something else. With the right level of faith by the hypnotist, one can make miracles.

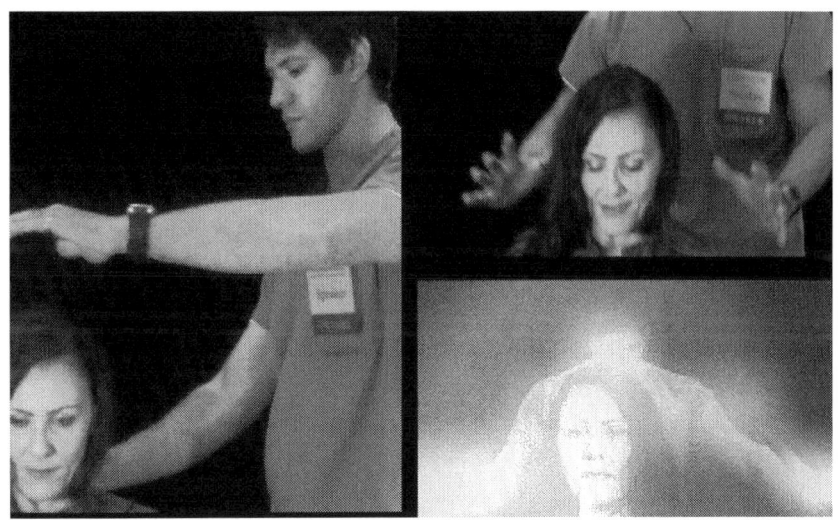

Chi Treatments

Chi treatments are a little different in that instead of allowing easier healing from relaxation, the healer uses a strong "faith" in his training. Very few peokle have gotten this CHI which essentially is a focused energy in the body. By meditation, diet, and some other secret things, masters of Chi can, somehow, heal people by directing this internal energy into a wound or disorder to modify it. I know it sounds like faith healing, but Orientals always want their own name.

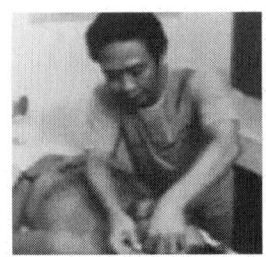  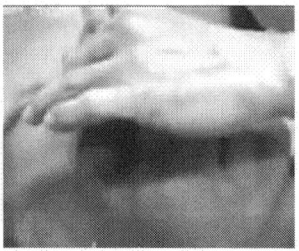

A demonstration of energy focus is shown above with extended accupuncture treatments, making flames appear and simply touching someone to cure issues. This is no joke. It seems to work, but hopefully, you see the hands on the person is, again a real part of this treatment. The focusing of energy may also give us a clue that large amounts of E-M fields that can be created by the Heart-brain.. The "prana" of Indian Yoga physiology apparently is similar to this Chi energy concentration and their hands can heal as well as faith healers in India want their own name as well.

Body Psychotherapy

Just like treating disease, laying on of hands seems to work for depression and other psychological disorders. You might see a body psychotherapist if you've got a persistent case of the blues; to help you cope during a troubled time, such as a death in your family; or to deal with the effects of a past trauma, like being the victim of a crime. The patient simply lies down while a therapist <u>touches</u> different spots on the body. In one account, a therapist simply <u>touched the back of the neck</u> of a patient and it triggered a terrifying memory. Through the hand touching as allowing a memory to be revealed reduces the trauma of the patient. It is like having a psychologist with fingers. Of course, this again is the Heart-brain E-M field at work.

Faith-Healing

In actual fact, in so far as faith-healing is concerned, religion is not all that important. There are numerous cases of faith-healers performing their faith-healing acts without using religion at all. Cases on this point are the healing methods I presented previously and the Bible indicates that <u>many</u> faith healers would be going to Hell. Here are a few examples.

Matthew 7:22*-Many will say to Me on that day, 'Lord, Lord, did we not prophesy in Your name, and in Your name* [saying Jesus in no way means someone is a follower] *drive out demons and perform many miracles?' Then I will tell them plainly, '<u>I never knew you; depart from Me, you workers of lawlessness</u>.'* [Many unbelievers were able to cure sick, work miracle, and drive out demons]

Luke 9:1 *When Jesus had called the <u>Twelve</u>* [Judas and the other 11] *together, he gave them power and authority to drive out all demons and to cure diseases.* [This included the disciple called unclean, Judas Iscariot. While he was not a true follower, he still cured disease and forced demons out of people using "faith".]

Luke 9:49 *John answered and said, "Master, we saw someone <u>casting out demons</u> -; and we tried to prevent him because <u>he does not follow along with us</u>."* [This unbeliever cast our demons by faith]

***Acts 19:13**-<u>Some Jews who went around driving out evil spirits</u> tried to invoke the name of the Lord Jesus over those who were demon-possessed. They would say, "In the name of the Jesus whom Paul preaches, I command you to come out." Seven sons of Sceva, a Jewish chief priest, were doing this.* [One time it didn't work.]---*they ran out of the house naked and bleeding.* [We don't know how many people this group of unbelievers helped, but it was not always successful.]

Again, all that is necessary for focusing you heart-brain to help effect a cure for disease is this faith stuff.

Just like in the Bible, there are many discourses in the "Teaching of the Buddha" where it was indicated that various forms of sicknesses were eradicated through the "conditioning of the mind; thus, it is worthwhile to practice meditation in order to attain mental and physical well-being or help others with their sicknesses. Instead of using faith, they would advance their Chakra level. We will look at faith and Chakra level later to see how this can amplify the capabilities of the Heart-Brain. First, here are a few of the well witnessed "healings" using the focused hand method.

Ruptured Disk Cure-The patient had a ruptured disk that caused him unending pain. A co-worker performed hands on healing which caused a feeling of great heat. The pain left and returned 5 times before finally having ALL PAIN completely removed. After returning to the doctor it was determined that the ruptured disk had <u>miraculously</u> healed!

Lost Leg Pain-This same co-worker helped another who had lost a leg in the Vietnam War. The healer prayed and laid on his hands and in 5 minutes the hurting area was very hot. After one treatment, pain that had been had for 20 years was gone.

Skin Cancer-A melanoma was eliminated within one day by this same faith healer and was a visible testimony of the success of this work.

Cat Healing-No Placebo effect for sure here as a stupid cat was cured in one hands-on session that was almost certainly going to die.

Traiteur Healing-It is common in Louisiana Cajun culture to have persons who have the gift of healing, thought to be a blessing from God, but those who practiced this ancient "ART" seemed to have been too young to understand what they were doing when they started healing. These special people are known as *traiteurs*, [Similar to the 'powow' healer of the Pennsylvania German community or the 'power doctor' in the Ozarks] They do not advertise their powers and never take money.

Buddha Healing-In Tibetan Buddhism faith healers are said to be given the "gift" from Buddha. About the same as the Christian healing, but again, this belief "faith" was developed by advancing something called Chakra awareness.

How Do You Get Faith?

Let me first start with the Biblical instruction. Here we find that the way to faith is to lose our connection with our "self". It's going to die one day anyway. Why shouldn't we concentrate on our other 3 entities? The New testaments tells us the same thing over and over and over. While this not only builds faith, it also allows for later salvation, but that isn't part of this book. I simply want to tell you how to use your Heart-Brain.

Matthew 10:39- *He who <u>finds his life will lose it</u>, and he who loses his life for My sake will find it. Accept what you are able to do and what you are not able to do.*

Matthew 16:25- *For whosoever will <u>save his life shall lose it</u>: and whosoever will lose his life for my sake shall find it.*

Mark 8:35- *For whosoever will <u>save his life shall lose it</u>; but whosoever shall lose his life for my sake and the gospel's, the same shall save it.*

Luke 9:24- *For whosoever will <u>save his life shall lose it</u>: but whosoever will lose his life for my sake, the same shall save it. For what is a man advantaged, if he gain the whole world, and lose himself.*

Luke 17:33 *-Whosoever shall seek to <u>save his life shall lose it</u>; and whosoever shall lose his life shall preserve it.*

John 12:25- *He that <u>loveth his life shall lose it</u>; and he that hateth his life in this world shall keep it unto life eternal.*

Thomas 1:56 *- Jesus said, "Whoever has come to <u>understand the world has found only a corpse</u>, and whoever has found a corpse is superior to the world."*

This same thing is described in other New Testament texts as well, but it is repeated over and over as it is one of the most important things for us to learn so that we can become happy and, at the same time free up our heat brain. Forget yourself and you will automatically enhance the capability of you Heart-Brain. Certainly meditation, prayer, and fasting allows one to get to a "better place. Some use drugs to eliminate the desires of the flesh, others have used an Infratonic machine, or a transcranial Magnetic stimulator. Without meditation and focus away from self, there is no hope. Just keep believing--- *If you want to live* [and expand the capabilities of the Heart] *you must die* [at least the "self" portion of you and carnality must be eliminated.] While I indicated faith in God was not needed for this faith healing capability, it really would be helpful as that by itself reduces your constant adoration of self.

Latter we will get to definitions called Chakras that will help us understand these types of communication levels and what is needed to advance them. Right now, let's look at the Heart and how it controls emotions.

Heart and Emotions

As we find in a number of the Biblical descriptions, the heart is intimately needed to support emotions. Now we know that it absolutely is the master of hormone balance, emotional well-being, and calming to all our Head- brain to think properly.

This description is fairly well recognized and described in the report by *[Rein, Atkinson, et al, 1995]-It is long known that changes in emotions are accompanied by predictable changes in the heart rate, blood pressure, respiration and digestion. So, when we are aroused, the sympathetic division of the autonomic nervous system energizes us for fight or flight, and in more quiet times, the parasympathetic component cools us down. In this view, it was assumed that the autonomic nervous system and the physiological responses moved in concert with the brain's response to a given stimulus.*

What we will find as we go along is that the Heart-brain not only controls the emotional elements of our bodies, but it also sends massages outside the body to affect those around us. These messages include the emotion of fear, anxiety, love, trust, friendship, caring, oneness and other similar emotion centric messages. This can only be done with a secondary consciousness of our Heart outside the brain.

Heart Consciousness

As I mentioned, the heart brain, may well be more conscious of the "environment than our head-Brain. Certainly, the heart is more aware of the soul and spirit as it interfaces with both according to the Biblical testimony.

This is because, the Heart not only filters information from our consciousness, but also from our free-floating Soul and the intra-universe communicating spirit [This part of our existence sort of to Heaven and back]. Additionally, it seems the intimate association of feelings appears to be supported much more by the Heart brains than our head brain. There is a verse in Proverbs written by a man that was supposedly the wisest man, ever. Solomon's words tell us what we will learn in this book.

Proverbs 4:23 *<u>Above all else, guard your heart, for everything you do flows from it.</u>*

Introspective Awareness

There is now an idea of "<u>introspective awareness</u>" of the heart itself. Possibly you can say we are conversing when only you, your head brain and your heart brain are in the room. While this is the scientific rage, I am more inclined to think that the <u>soul uses the heart-brain to communicate with others</u>. As we remove our self-centered emotion to the emotion of love and protection of others, our soul becomes very powerful [increase primary faith] according to Biblical teachings. This will make more sense as we go along.

Heart Sort of Sees the Future

I was going to tell you about another study that showed the Heart brain could understand stimuli faster than the head brain, but that is starting to sound twilight zone-ish. Oh well, I might as well tell you now. The heart can provide a split-second "body premonition." Essentially, test subject's hearts "felt" the future before it happened. Don't ask me to explain this at all as we continue. This is simply science. As the heart senses the world, communicates with the Head and outside the body, and it established the control of the body using blood pressure and hormonal discharges, it learns its environment in some ways with more detail that does the Head-brain.

Heart Brain Learning

As I mentioned, our Heart Brains are somehow directly tied to our Souls. What we are finding is that even after death those "learned" elements are carried forward to our next life or to our Heavenly destination. What I mean is the Head-brain information is lost at death, but the Heart-brain hold memories in some special ways and transfers feeling, love compassion, trust and other elements of our lives to our freed souls when "Body death" occurs. While we are "Alive" the Heart -brain is at work, checking out the environment and processing the information it obtains.

Heart Information Processing

Research in the past two decades has shown that the heart is an information processing center that can learn, remember, and act independently of the cranial brain and actually connect and send signals to key brain areas such as the amygdala, thalamus, and hypothalamus, which regulate our perceptions and emotions. Once a partner is learned by the heart it seems the levels of love based hormones and the "coherence heart rate frequencies" come about much more often when those people are nearby. This simply means that when two lovers are together, their hearts and minds begin functioning in a level of synchronization. It seems we can only second "brain" in our chest. While some of this memory might be pheromone based, it is looking more and more like Heart brains communicate with each other over their E-M communication channel. If we find out how to talk to it we may even cure lovesickness.

If you want your body to feel better and get better, being in a room full of loved ones certainly helps. If you are in a coma, loved ones may be able to communicate with your heart.

This is not removed when someone is in a coma. Indications show, loved ones visiting coma patients assure a more rapid recovery and their heart-brain communications continued even if the brain functions were limited for a time. If your friend is in a coma, go visit and tell him or her, you care about them. Better yet; have your heart-brain tell them. With all this heart-brain control what happens if you get someone else's heart?

Heart-Brain Control

What does the heart-brain do with all its memory, control of hormones, internal and external electro-magnetic communication and the like? Researchers are finding is that the heart brain not only stores memories, it also takes a major place in making your body and mind work properly.

The ability of the brain, through the and nervous system, to synchronize electrical activity is the basis of what we call <u>convergence of consciousness</u>, our ability to perceive, feel, focus, learn, reason and perform <u>at our best</u>. Disruption of this system can be causes by stress. It is well established that stress interferes with mental processes such as memory, concentration, judgments and decision making. There is more information going from the heart to the brain than the other way around, and this information influences region's in the brain that affect decision making, creativity and especially emotions.

Heart Brain does most of the communicating

I know you are thinking that with our massive brain, the body is sort of a downward system where the head-brain commands actions and the body does them, but instead we find that 85-90% of all neural fibers carry information from the body to the brain, and a major part of this information comes from the heart via the nervus vagus.

Certainly, the Head-Brain communicates with the heart, but reaction time is problematic, therefore, the Heart uses its own brain to speed up reaction time and something quite different. We are now finding that the Heart-brain communicates with the Head-brain much more often that the other way around. This communicating controls many aspect s of our lives, but the most important is the control of stress which affects almost all other parts of brain function. Research shows that the heart communicates to the brain in four major ways:

- Neurological communication **(nerve impulses)**
- Biochemical communication **(hormone release)**
- Biophysical communication **(pulse waves)**
- Energetic communication **(electromagnetic fields)**

The research of these communications has been going on for about 50 years now. It is a slow process, but every year we find out more.

More Active Than the Head Brain-According the McCraty and the Institute of HeartMath, *the heart is in a constant two-way dialogue with the brain and the heart system is sending far more signals to the brain than the brain is sending to the heart. At the same time, the heart-brain has direct connections to organs such as the lungs and esophagus and are also indirectly connected via the spinal cord to many other organs, including the skin and arteries, so its direct neuronic messaging is substantial. Its hormonal messaging affects many body functions, its E-M communication link is 5000 times as strong as the Head-Brain, it feels things for those around you, and now we believe it remembers what its likes and dislikes were. This* last thing has cause issues with heart transplant recipients, but I'll have to get to that later.

If you want to stay healthy you had better get a compatible mate. I don't mean someone your head-brain thinks is sexy.

Somehow, we need to ask our heart who we should be with. One can make the connection that the Soul gateway to the body is through this special brain rather than the one on top of your head. A brought-up coherence earlier, but let's describe it as the equal discourse between the Head-brain, Spirit, Soul and Heart-brain as shown below with the heart being in the very center of everything. Anytime, the heart has to pay more attentions to other segments of our being, things change.

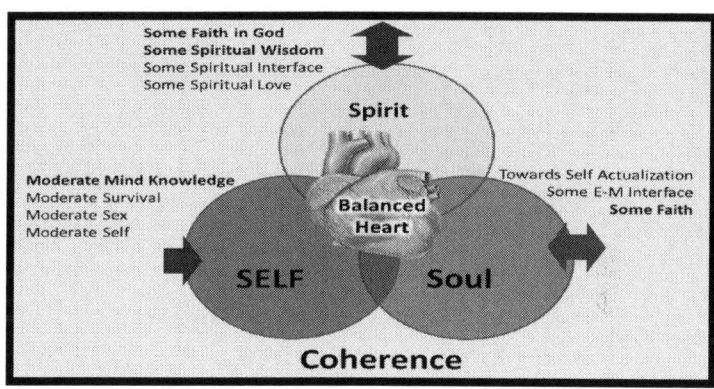

Let's see what our Bible had to say.

Job 10:13 *What is <u>concealed in your heart, and I know that this was in your mind:</u>*

Deuteronomy 4:29 *But if from there you seek the Lord your God, you will find him if you <u>seek him with all your heart and with all your soul.</u>* This verse tries to tell us about how the heart and soul are closely connected.

Job 19:27 *How my <u>heart yearns</u> within me!*

***2 Corinthians 4:16**-Therefore we <u>do not lose heart</u>, but though <u>our outer man is decaying</u>, yet our inner <u>man is being renewed</u> day by day.*

***Matthew 22:37** Jesus replied: <u>'Love the Lord your God with all your heart and with all your soul and</u> with <u>all your mind.</u>'* While this sounds like the Heart is associated with the Spirit, the third part of our being, it is really reaffirming the description of Deuteronomy.

> ***Jeremiah 29:13** You will seek me and find me when you <u>seek me with all your heart</u>.* This is firmly stating trying to understand God with only your head brain is impossible.

***Proverbs 3:5,7** Trust in the Lord with all your heart, and lean not on your own understanding.... Do not be wise in your own eyes.* [False wisdom is obtained by the Head-brain, true wisdom comes from the Spirit.]

It was as if the Bible was way ahead of these scientists, but now we are beginning to see the heart is an important gateway between the self, the soul, and the spirit. Keeping our triune entity working together is a great thing for our "Self" as it regulates hormones, synchronized the head-brain and reduces stress. Pioneers in the field recognized the importance of this link and organized international conferences on the Heart & Brain. This community includes fields of neurologists and cardiologists. The cardiologists describe the brain and the neurologist learn how the heart is affects the brain. Together they are beginning to peel back the secrets of the Bible and understand the remarkable regulation of our heart. That being said, just staying in coherence with you own head-brain may not always be best as the Heart-brain can communication outside the body.

Communicating Heart Brain

Not only have we found the heart-brain communicates with those around it, it can sort of tell adjacent Heart-brains to act in a similar way. The way this is done is the same as normal Electro-magnetic [EM] communication like radio waves that somehow change the receiving radio by its signals.

The heart-brain radiates a <u>rhythmic EM field</u> with every heartbeat that affects those in our environment, whether we are conscious of it or not. The rhythm of those around a transmitting Heart-brain attune to the sender.

If a sender if emotionally disturbed, those nearby will become disturbed. If the heart-brain is mellow, others become mellower. If the heart-brain exudes love, there will be a more prevalent feeling of love. Researchers call this coherence as it links the soul, spirit, head-brain and heart together.

This electromagnetic field of the heart-brain is 5000 times stronger than that of the head-brain, and generally measurable up to 3 feet away from the body.

Here is an example: When a mother has a feeling of joy while holding her baby and the baby has nothing to distract it, the pulse rates of both seem to "synchronize."

One experiment had this "synchronization" effect occur even when subjects were as far apart as 5 feet. In another test, the heart rates of a horse and rider would synchronize to some level.

When we focus more on our soul, we communicate with and through it. You have seen in the movies where someone telepathically tells another person to do something and they do it. Unfortunately, the heads brain capability of transmitting Electo-magnetic messages to another person is very weak, but the Heart-brain E-M transmissions are strong, low frequency magnetic pulsing waves. We are told, in a number of ancient histories, that prior to 3100 BC [Tower of Babel or Bharata War timing], the head-brain had a stronger transmission capability and we could even communicate with others without speech, but now only our heart –brain can successfully communicate with those nearby, without speech.

The heart-brain "circuitry" enables it to sense, regulate, and remember a vast array of emotions, action, stressers, and many other things we are just now starting to understand. This information is used not only regulate and control many functions in our own body, but also is looks like the Heart-brain talks to others nearby, sends out these findings and remembered events and modifies the stored and used events of those Heart-brains and/or receives signals that make changes in our Heart-brain and its memory. Let me tell you what that is called ---COMMUNICATION.

Soul Focused Heart-Brain

Several passages in our Bible describe the Heart-brain working with the Soul. It appears to be the channel that allows the Soul to expand the awareness of reality beyond what we can see feel or hear and something that causes faith. I'm not talking about faith in God, here. I'm talking about faith to cure sickness, walk on water, and even bring people back to life. We can call Faith in God **"Faith#1"** and this other thing **"Faith#2"**. Unfortunately, both are called out by the same word in our Bible so we are kind of stuck with it in this type of discussion, but I will show some of the defining verses later. To gain this Faith #2 you must be attuned to your Soul through your heart-brain. Here are some verses that help us understand.

What is Faith #2?

While there are 2 types of faith described in the Bible [Faith in God for Salvation and Faith like a mustard-seed to perfume miraculous deed-healing the sick, walking on water, turning water to wine and others], these verses may hold the key to faith needed for healing the sick. [We will get to the other faith later.] We are told in the Bible that many unbelievers had faith enough to heal the sick, preach insightful messages, and turn water to wine or similar changes in this reality. Possibly, the way this whole thing works is that the Soul and Heart-brain typically work in concert, but most of the time, that portion of communications are locked from our conscious mind. All that is needed is for us to eliminate our hardened heart-brain so that we can interface with the Soul and we are stuck with the self-centered, evil thinking, sex driven, and survival focused head brain.

The Bible certainly tells us this openness or faith is not enough to provide salvation which must add in direct unrestricted communication with the spirit through the heart-brain.

Luke 17:6 *And the Lord answered, "If you have faith [#2] the size of a mustard seed, you can say to this mulberry tree, 'Be uprooted and planted in the sea,' and it will obey you.* [Nothing to do with salvation]

Matthew 17:20 *"Because you have so little faith." He answered. "For truly I tell you, if you have faith [#2] the size of a mustard seed, you can say to this mountain, 'Move from here to there,' and it will move. <u>Nothing will be impossible for you</u>.* [Nothing to do with salvation]

Matthew 7:22 *<u>Many will say</u> to Me on that day, 'Lord, Lord, did we not <u>prophesy</u> - <u>drive out demons</u> and <u>perform many miracles</u>?' Then I will tell them plainly, 'I never knew you; depart from Me, you workers of lawlessness.'* [These people clearly had Faith #2 and no Faith #1.]

Hebrews 4:12 *the word of God penetrates even to dividing* **soul** *and* **spirit***; it judges the thoughts and attitudes of the* **heart.** [Just because your heart-brain has substantial control and interface with the soul, does not mean you have faith #1.]

Luke 9:49 *John answered and said, "Master, we saw someone <u>casting out demons</u> -; and we tried to prevent him* *because he does not follow along with us."* [Unbelievers cast our demons by faith]

Hosea 11:8 *My heart is changed within me; all my compassion is aroused.* [Compassion for people around you is form the Heart communicating with the soul.]

Matthew 11:29 *I am gentle and humble in heart, and you will find rest for your souls.* [True humility, is from the Heart communicating with the soul.]

Acts 2:26 *my heart is glad and my tongue rejoices* [Being truly glad for people around you is from the Heart communicating with the soul.]

2 Corinthians 9:7 *Give what you have decided in your heart to give* [Compassion for people around you is from the Heart communicating with the soul.]

Colossians 3:22 *Slaves, obey your earthly masters with sincerity of heart* [Sincerity with those around us is from the Heart communicating with the soul.]

Today we find many faith healers using "Faith #2", and the power of a focused soul through the heart-brain. If we wanted to look at the focus of the soul centered Heart-brain it would look something like this.

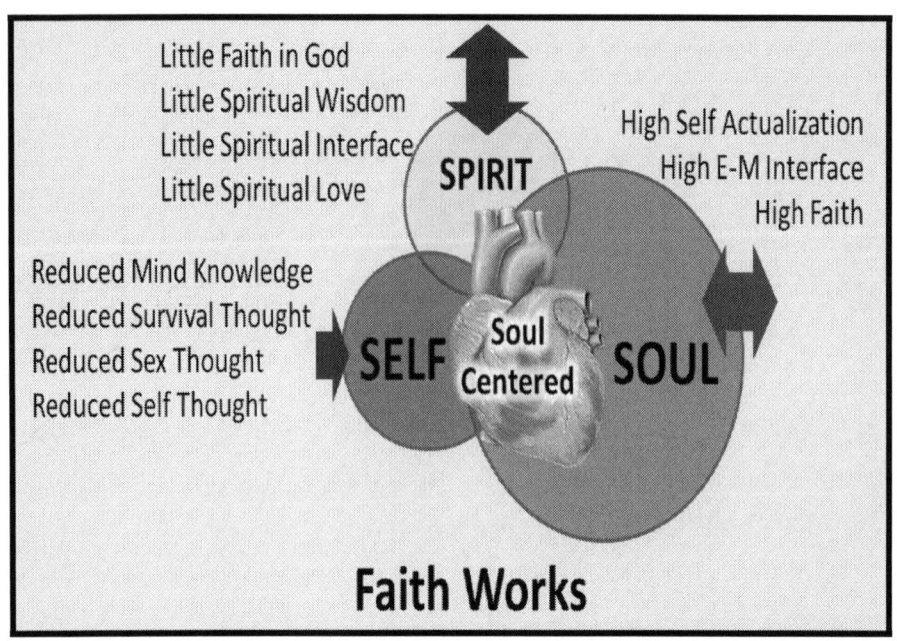

Faith Works

These are not bad people nor is it bad to focus the Soul to engage outside the body and use the E-M communication capability of the Heart-brain.

Faith #1

If all the soul faith is being directed towards serving the Creator, faith become Faith #1 as described below.

Deuteronomy 6:5 *Love the LORD your **God** with all your **heart** and with all your **soul** and with all your **strength**.*

Deuteronomy 10:12 *fear the LORD your God, to walk in obedience to him, to love him, to serve the LORD your **God** with all your **heart** and with all your **soul***

Deuteronomy 13:3 *love him **[God]** with all your **heart** and with all your **soul**.*

Deuteronomy 30:6 *The LORD will circumcise your hearts so that you may love him **[God]** with all your **heart** and with all your **soul**, and live.*

Matthew 22:37 *Love the Lord your **God** with all your* <u>**heart** and with all your **soul** and with all your **mind**.</u>

Mark 12:29 *love the Lord your **God** with all* <u>your **heart** and with all your **soul** and with all **your mind** and with all your **strength**.</u>'

Mark 12:33 *To* <u>love him with all your **heart**</u>*, with all your **understanding** and with all your **strength**,*

Joshua 22:5**-walk in obedience to him, to keep his commands, to hold fast to him **[God]** and to serve him with all your **heart <u>*and with all your **soul**.*</u>

Heart-Brain Sight

Some of the verses about the Heart appear to describe the heart seeing and it absolutely does. What scientists have found is that that sees at a much lower wavelength than our Eyes, but it does sense electro-magnetic [E-M] waves just like our eyes. This is an exciting component of our Heart brain and it associated efforts. What the Heart sees is stress, love, hope, understanding and tension from the Heart brains of other individuals nearby. Our heart sends out messages to calm of stress the environment and it receives this same indication from others nearby. It has been long known that if you place individuals together if more are hostile, soon almost all with take on that feeling and vice versa. A calm or pleasant emission will make other conform to that calmness or goodness. This is a great responsibility. Smile at those you pass by as smiling helps trigger he good E-M transmissions.

The Heart's Electro-Magnetic Field

Research has also revealed that the heart communicates information to the brain and throughout the body via electromagnetic field interactions.

The heart generates the body's most powerful and most extensive rhythmic electromagnetic field. The heart's magnetic component is about <u>5000 times stronger</u> than the brain's magnetic field and <u>can be detected several feet away from the body.</u>

It was proposed that; this heart field acts as a carrier wave for information that provides a global synchronizing signal for the entire body.

Heart Interactions Between Individuals

Additional neurophysiologists including Dr. *McCraty [2002], jumped on the bandwagon. He discovered a neural pathway and mechanism whereby input from the heart to the brain could inhibit or facilitate the brain's electrical activity.*

There is now evidence that a subtle yet influential electromagnetic or 'energetic' communication system operates just below our conscious awareness. Energetic interactions possibly contribute to the <u>'magnetic' attractions or repulsions that occur between individuals</u>, and also affect social relationships. It was also found that <u>one person's brain waves can synchronize to another person's heart</u> according to work done by McCraty in 2004.

Now think about this for a minute assuming the heart as a tool for the soul. As we give our soul more allowance to "wander" this communication with others can allow us to not only affect our own reality, but also change others through this rudimentary communication link. Again, when I'm talking about soul wander I'm talking about the soul [reality] in combination with the spirit [anti-carnality].

105

Ephesians 1:18 *I pray that the* <u>**eyes of your heart** may be enlightened</u>

When we look at Proverbs 15:30 we see that if someone comes near another's Heart-brain, the brain can sense/see if the person is filled with the spirit/light or not.

Proverbs 15:30 *Light in a messenger's eyes brings <u>joy to the heart</u>,*

Intuitive Heart-brain

While Jeremiah talked about the yearnings of the heart-brain as if the heart would preconceive emotional characterizations before the Head –Brain could pick up of some emotion established from an event. Researchers found out that is exactly what happens.

The Heart-brain seems to know an emotional event is "ABOUT" to happen.

Here is how the testing was set up. A researcher and a student would sit at a table and the researcher would show some event, horrific destruction, viewing over the edge of a deep canyon, having a baby and the like. These were mixed in with less emotional images flying kite, clouds, hand shaking etc. The researcher and the student blood pressures were monitored for rate and disorganized fluctuation. The researcher's blood pressure indicators would take longer to show emotional response than those of the student that viewed the same images a few seconds after the researcher.

Think about this for a minute. Either the heart-brain was anticipating the events or there was a communication of emotion from the researcher to the heart-brain of the student that was picked up before the researcher's brain even knew he was being affected.

There is a significant deviance in the heart rate variability when an emotional stimulus is about to be shown to the student about 4.5 seconds before it was actually shown. This variance is more noticed when the person being tested is emotionally "happy" than "stressed" and women showed a much stronger effect than the men in being attuned to their heart.

The old adage women are more intuitive than men may be associated with the intuition of their heart-brain.

If the heart is not attuned to the spirit, or if the spirit is not "released" to communicate, it causes something called Hardening or callusing of the Heart-brain.

Hardened Heart-brain

This is one of the more difficult areas of the Heart-Brain description, but let me attempt to shed light on a very delicate description as stated in the Bible. First, we must concede that the Heart–Brain is needed for much more than pumping, blood, regulating mood, advancing dosages of hormones when needed, and adjusting blood pressure to regulate the body operations and all the rest and understand about how the Heart-brain interacts with the Spirit portion of our being. There is not a more mystic statement in the Bible than when the Bible states that "God hardened the heart". It appears 15 times in the Bible and 10 of them were to harden the heart of the Egyptian/Hyksos Pharaoh's heart [King Apepi] so he would only listen to his head brain and not allow the Jews to leave. This was said "to allow the wonders of God to be shown." That wasn't the only person as God hardened the heart and spirit of Sihon, king of Heshbon in Deuteronomy 2:30, and God hardened the hearts of the northern kings in Canaan in Joshua 11:20, the prophet Isaiah asks why God has hardened Israel's heart in Isaiah 6 and then again in chapter 63 as he was confused about this limitation of the Heart-brain. The apostle Paul stated that *"whom God wills He hardens"* in Romans 9:18 and he explained what was meant in Romans1:21-28. First let's review some of the important verses and go from there and explain this oddness.

Romans 1:21 - *Their thinking became futile and their <u>foolish hearts were hardened</u>. Therefore, God gave them over in the sinful desires of their hearts to sexual impurity for the degrading of their bodies with one another, --to shameful lusts. Even their women exchanged natural sexual relations for unnatural ones. In the same way, the men also abandoned natural relations with women and were inflamed with lust for one another. Men committed shameful acts with other men -- God gave them over to -- every kind of wickedness, evil, greed - envy, murder, strife, deceit and malice. They are gossips, slanderers, God-haters, insolent, arrogant and boastful; they invent ways of doing evil; they have no understanding, no fidelity, no love, no mercy.*

We will look at other descriptions, but Paul indicated a hardened heart would erupt when people continued to close off from the spirit portion of their entity. Essentially, unless someone uses his heart-brain to go outside his body, it will over time become more and more self-loathing, self-aggrandizing, and everything else "self" in between. Here are some more texts.

Isaiah 6:9-10 *Be ever hearing, but never understanding; be ever seeing, but never perceiving.' <u>Make the heart of this people hardened</u>; otherwise they might understand with their hearts, and turn and be healed."*

John 12:39-41 *Therefore they could not believe, because that Isaiah said again, <u>He hath hardened their heart</u>; that* *they should not understand with their heart, and be converted, and I should heal them.*

One portion of the reasoning for this whole Heart- brain hardening can be described next from Jeremiah's and Jesus' words.

Jeremiah 17:9 - *The heart is deceitful above all things, and desperately wicked: who can know it?*

Matthew 15:19 *out of the heart come evil thoughts—murder, adultery, sexual immorality, theft, false testimony, slander.*

These go along with the Romans texts. While the strange "hardening the Heart brain by God is initially disturbing, we also read that God can take away a hardened heart-brain to allow us to live more open to the world outside our "fixed reality".

Ezekiel 36:26 *I will give you a new heart and put a new spirit in you; I will remove from you your hardened heart and give you a heart of flesh.*

Don't think the ancient prophets had all the answers either. The great prophet Isaiah struggled with this whole hardening of the Heart brain just to show God's power.

Isaiah 6:10-14 *Rebekah's children were conceived at the same time - Yet, before the twins were born or had done anything good or bad——she was told, "The older will serve the younger." Just as it is written: "Jacob I loved, but Esau I hated."*

> *What then shall we say? Is God unjust? Not at all!*

Isaiah 6:15-19 *For he says to Moses, "I will have mercy on whom I have mercy and I will have compassion on whom*

I have compassion."- For Scripture says to Pharaoh: "I raised you up for this very purpose [to live through the plagues], that I might display my power in you and that my name might be proclaimed in all the earth." Therefore, God has mercy on whom he wants to have mercy, and he hardens [the heart] whom he wants to harden. One of you will say to me:

> *"Then why does God still blame us? For who is able to resist his will?"*

Isaiah 6:20-23 *But who are you, a human being, to talk back to God? "Shall what is formed say to the one who formed it, 'Why did you make me like this?' Does not the potter have the right to make out of the same lump of clay some pottery for special purposes and some for common use? What if God, although choosing to show his wrath and make his power known, bore with great patience the objects of his wrath—prepared for destruction?*

> *What if he did this to make the riches of his glory known to the objects of his mercy, whom he prepared in advance for glory?*

Isaiah continued searching for an answer and in chapter 63 we find this.

Isaiah 63:17 *Why, LORD, do you make us wander from your ways and <u>harden our hearts</u> so we do not revere you?*

For certain Isaiah knew that the Heart was more than a muscle pumping blood and it was intimately connected with our being able to understand things outside the "reality" that we see, touch, and feel. Intimately tied to both our soul and spirit. While there may be times when the Heart-brain and

112

the soul do not communicate, the most serious to our "afterlife" is how the Heart-brain works with the spirit. I explained how the spirit provides for 7 major character correctors needed to become more spiritual, what I did not discuss is how a person separating the Heart from the spirit has devastating consequence with respect to understanding God. If God showed his power and capability for devastation, we would certainly knuckle under and, for an instant, forget our carnal, lustful, hateful, self-centered, prideful, evil minded self like an ant viewing a massive foot about to pounce. Pulling his spirit away from our heart allows a person to act "normally" so God can enact awesome character and expand the understanding of his might, power, omniscience, omnipresence, and omnipotence----but that is not the fullness of cause for him hardening the communication between the Spirit and the Heart. The second part has to do with regeneration.

Hardening helps Define Reincarnation

As indicated throughout the New Testament; *God is not willing for anyone to perish in hell*. Therefore, he invented reincarnation and that will help us with this whole matter. While sometimes reincarnation is done to fulfill prophesy like when God Incarnate indicated *"Elijah came back as John the Baptist"*, but most generally, reincarnation is to allow a second chance for those wandering too far from God's ways. Hardening their hearts sometimes is used to shorten their current wayward life to allow their regeneration and possible acceptance in another. Other times, God told his people to eliminate the Amalekite race so that they would have a second chance. God uses hurricanes, natural disaster and even the harden hearts of angry people to shorten lives for reincarnation and possible redemption.

2 Peter 3:8-9 *With the Lord a day is like a thousand years, and a thousand years are like a day. The Lord is not slow to fulfill His promise as some understand slowness, but is patient with you, <u>not wanting anyone to perish, but everyone to come to repentance</u>.* [While this initially doesn't make sense that God would harden anyone's heart until we read further.]

Matthew 11:14 *The Law prophesied until John and if you are willing to accept it, <u>he [John the Baptist] is the Elijah who was to come</u>.* [Actually, Elijah never technically died as his soul was taken from his live body, but this absolutely indicates that Jesus knew Elijah's soul had entered into a new body to become John the Baptist as a reincarnation that was considered normal during Jesus time and should be recognized today as well. Otherwise God Hardening someone's heart will not make sense.]

Deuteronomy 7:1-2 *–[Talking to the Jews] nations before you, the Hittites, the Girgashites, the Amorites, the Canaanites, the Perizzites, the Hivites, and the Jebusites, -- when the LORD your God gives them over to you,- <u>you must utterly destroy them.</u>* [Like hardening one's heart, this initially seems cruel, but it was to give them a chance at salvation. If some coming back were in situations allowing them to follow the Incarnate God they would be a thousand times better off.]

Romans 8:28 *And we know that <u>in all things</u> [hardened hearts, floods, fires, war, etc.] <u>God works for the good</u> of those who love him.* [This initially seems like hardening one's heart, but it also is to provide mercy by reincarnation.

Yes, some hardships, deaths of love ones, sicknesses and the like are to "train" a believer, but many who die in disasters are not followers so their death would have no meaning without reincarnation.]

Some Will Not Sleep In Death

The next few verses even had the Apostles thinking Jesus would come back in their lifetimes, but he still has not returned 2000 year later. The last one is the most telling so let's look at that one specifically as all of the disciples had their bodies die.

John 8:51-Truly, truly, I tell you, if anyone keeps My word, he <u>will never see"</u> death.

Matthew 16:28-Truly I tell you, <u>some who are standing here will not taste death</u> until they see the Son of Man coming in His kingdom."

1 Corinthians 15:42,43,51,52- Our bodies will be raised "in incorruption, glory and power"--- We shall <u>not all sleep</u>, but we shall all be changed. In a moment, in the twinkling of an eye, at the last trump: for the trumpet shall sound, and the dead shall be raised incorruptible, and we shall be changed

When Paul described this no death thing, he changed it to "being asleep in death". Certainly, most people simply sleep and wake up to punishment or glory, but if a soul is not sleeping, he can return and become a reincarnated person to fix something or do what he should have done before. Evidently, some of the disciples have had many lives this way such that they did not sleep. As God had allowed many to be killed, he allowed them to reincarnate and have a better chance at allowing the heart brain to work its magic.

Calloused and Evil Heart

As described in other verses. A hardened heart is sometimes referred to as a callous or evil heart. This is because, without interference and yearnings of the spirit, the self-portion of our being will get more important to us and so both the Soul and the Spirit struggle for interaction. Without guidance of the Soul, one disregards what is the common good, while disregarding the promptings of the Spirit will certainly shift focus of one's life from receiving information from outside to only listening to the ravings of our own mind. The Heart brain becomes relegated to internal pressures of hormones rather than acting on relationships or anything outside the carnal mind. The person would become withdrawn or sociopathic. The diagram below shows the bloated self-portion making an incomplete person.

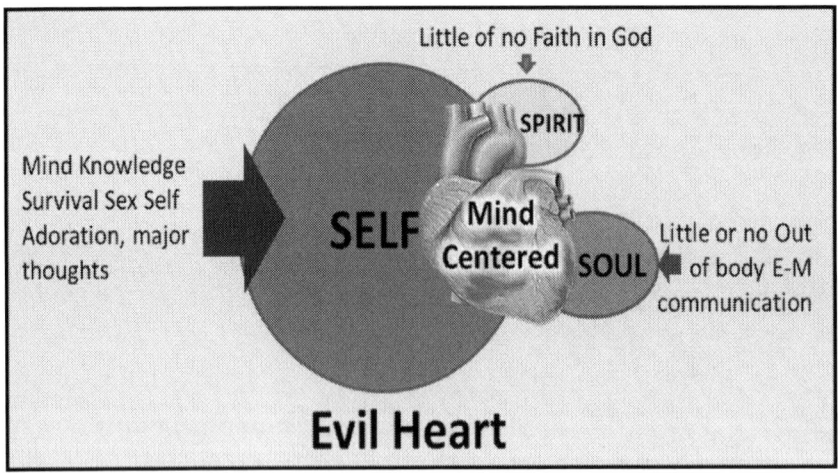

Hebrews 3:12 none of you has <u>a sinful, unbelieving heart</u> that turns away from the living God.

Hebrews 4:12 *the word of God penetrates even to dividing soul and spirit; it judges the <u>thoughts and attitudes of the heart.</u>*

Matthew 15:19 *out of the <u>heart come evil thoughts— murder, adultery, sexual immorality, theft, false testimony, slander.</u>*

Matthew 15:18 *<u>things that come out of a person's mouth come from the heart</u>, and these defile them.*

Matthew 12:34 *how can <u>you who are evil say anything good? For the mouth speaks what the heart is full of.</u>*

Matthew 13:15 *For this <u>people's heart has become calloused; otherwise they might understand with their hearts</u> and turn, and I would heal them.'*

Deuteronomy 8:14 *then your <u>heart will become proud</u> and you will forget the LORD*

Deuteronomy 29:18 *Make sure there is no man whose <u>heart turns away</u> from the LORD*

Job 36:13 *The <u>godless in heart</u> harbor resentment;*

2 Samuel 6:16 *Michal <u>despised him [David] in her heart.</u>*

Jeremiah 17:9 *- The <u>heart is deceitful above all things</u>, and <u>desperately wicked</u>: who can know it?*

Jeremiah 49:16 *The terror you inspire and <u>the pride of your heart have deceived you</u>,*

Psalm 58:2 *<u>in your heart you devise injustice</u>,*

Proverbs 6:25 *<u>Do not lust in your heart</u> after her beauty or let her captivate you with her eyes.*

Proverbs 12:25 *<u>Anxiety weighs down the heart</u>,*

2 Chronicles 32:25 *Hezekiah's <u>heart was proud</u> and he did not respond to kindness therefore the LORD's wrath was on him.*

2 Chronicles 32:26 *Hezekiah repented of the <u>pride of his heart.</u>*

Ezekiel 25:6 <u>*rejoicing with all the malice of your heart*</u>

Isaiah 13:7 *every <u>heart will melt with fear</u>.*

Hosea 10:2 *Their <u>heart is deceitful</u>*

Matthew 5:28 *anyone who looks at a woman lustfully has already <u>committed adultery with her in his heart</u>.*

Hopefully you can see the NORMAL condition of the heart is almost useless with regards to many of the things we are discussing in this book as our self-interest overshadows everything. Being out of touch with how one fits into this universe properly one might get feelings from the heart and head-brain of remorse, sadness, hostility, contempt, lust, egocentricity, money hording, contempt, jealousy, envy, greed, power, and hate. Swift mood changes would show up in this unhealthy condition. The problem is, many are stuck in this impossible life by not trying to reduce thoughts of self and begin to feed the other parts of one's existence.

There is no question that Jeremiah was right saying man has an evil heart. The apostle Paul simply said, man thinks of evil always. He could have said most men have an evil Heart-brain. [unfed, unused, atrophied.]

Heart Replacement Caution

The heart contains a complex intrinsic nervous system comprised of multiple clusters of neurons that network with each other and acts on their own. Unbelievably, <u>replacement of a heart is not devastating to the new owner</u>. One would think he would not get feelings of love, have the proper hormones released and not be able to tell the head brain what it needs to, but the E-M communication network of the Heart being 5000 times as strong as the E-M output of the head brain allows a stranger's heart to work in a new body. That being said there may be noticeable differences as the Heart-brain has its own likes and dislikes concerning people.

Heart Controls Love-Here is a scary thought. The type of people you liked before you gained a new hear might not be as compatible with your new heart as you would like. You may have to change significant others. It seems that a substantial proportion of people who have had transplants take on the likes, dislikes and certain characteristics of the heart donor. Scientists began looking into this strange occurrence. Depending on the researcher's subjects it has been found that between 6% and 23% of the recipients of heart transplants take on characteristics of their donor.

I should say that a second time and do one of those long pauses to let it sink in, but I still have too much data to tell you. What I will do is give you a solid number. There are about 4 thousand heart transplants done every year so we are talking about a huge number of Heart-brain memories transferred. Some of the examples can be found in the book "A Change of Heart" and others are from multiple eyewitnesses who saw, felt, or lived the reality of personality traits transferring to another body and I am not talking about minor changes.

Memory of Sex-One woman said she stopped wanting to have sex with her husband after concluding that her donor must have been a man.

Memory of Feelings- In 2008 shortly after receiving a heart transplant, Sonny Graham tracked down the wife of the donor – and fell instantly in love with her. "When I first met her," Sonny told a local newspaper, "I just stared. I felt like I had known her for years. I couldn't keep my eyes off her." He spoke of a deep and profound love for her. It was instant and it was passionate. The kind of love where overwhelming passion seizes control of the mind and banishes reason. They quickly wed. Then this seemingly happily married 69-year-old man living in the U.S. state of Georgia, shot himself without warning. He had shown no previous signs of unhappiness, let alone depression. Oddly, Sonny had received a transplanted heart from a man who had also shot himself – in identical circumstances.

Memories of Being Killed-A young girl received a heart transplant. Unbeknownst to her, the donor is a ten-year-old murder victim. She immediately began having recurring nightmares about the murderer, whose identity she claims to 'know'. Her mother contacted an initially-skeptical psychiatrist, who became disturbed enough to call in the police. Using the girl's descriptions, they pin-pointed the murderer. "The weapon, the place, the clothes he wore, everything was completely accurate."

Shot in Face Memory-A lecturer received the heart of a policeman who has been shot in the face by a suspect who 'looks like Jesus.' Three years later, the recipient confesses he is plagued by waking dreams: "A glimpse of Jesus, then a flash of light, then my face gets so hot, it actually burns."

Song Memory-An 18-year-old boy who wrote poetry, played music and composed songs was killed in a car crash. A year after he died, his parents came across a tape of a song he had written, entitled, Danny, My Heart Is Yours. In his haunting lyrics, the boy sang about how he felt destined to die and donate his heart. After his death, his heart was transplanted into an 18-year-old girl – named Danielle. When the boy's parents met Danielle, they played some of his music and she, despite never having heard the song before, knew the words and was able to complete the lyrics.

Memory of Daddy-A 7-month-old received the heart of a baby suffering mild cerebral palsy in his left side – then the baby developed shaking and stiffness on the left side. When the father of the donor saw the recipient, the baby called him "daddy".

Memory of Men-A vegetarian lesbian became 'man-crazy' after receiving the heart of a promiscuous straight woman. She also developed her donor's love of chicken-wings.

Memory of Music-A man received a young black violinist's heart. Afterwards, he had this to say: "I used to hate classical music, but now I love it."

Memory of Chicken Nuggets-On May 29, 1988, Claire Sylvia received a heart transplant. She was told that her donor was an eighteen year-old male from Maine. Soon after the operation, Sylvia declared that she felt like drinking beer, something she hadn't particularly been fond of. Later, she observed an uncontrollable urge to eat chicken nuggets and found herself drawn to visiting the popular chicken restaurant chain, *KFC*. She also began craving green peppers which she hadn't particularly liked before. Sylvia also began having recurring dreams about a mystery man named Tim L., whom she felt was the organ donor. She searched for obituaries and was able to identify the young man named Tim. After visiting Tim's family, she discovered that he used to love chicken nuggets, green peppers and beer.

Memory of Heterosexuality- The donor was a 19-year-old woman killed in an automobile accident. Her mother reported: "My Sara owned and operated her own health food restaurant and scolded me constantly about not being a vegetarian. She was into the free-love thing and had a different man in her life every few months. The recipient reported: "You can tell people, but it will make you sound crazy. First, almost every night, and still sometimes now, I actually feel the accident my donor had. I can feel the impact in my chest. Also, I hate meat now. I can't stand it. I was *McDonald's* biggest money-maker, and now meat makes me throw up. What really bothers me is that I'm engaged to be married. He's a great guy and we love each other. The sex is terrific. The problem is, I'm gay. At least, I thought I was. After my transplant, I'm not. Women still seem attractive to me, but my boyfriend turns me on; women don't."

Memory of Art-From the British tabloid, *"The Daily Mail"* we find that William Sheridan, had bad drawing skills and heart. His donor was an artist and William, all of a sudden, developed artistic talent.

Memory and Sadness-A man who was dying of bad lungs, received a heart and lung transplant from a young woman who had just died. Since his old heart was still robust, it was transplanted into another named Fred. After this transplant, Fred who was formerly laid-back began exhibiting the Type A aggressive behavior of the first man ad he began calling his wife, Sandy, the name of the original heart donor's wife. On the other hand, the first man became morose and sullen after the transplant. His donor had been a shy, soft-spoken young woman who had committed suicide.

Reincarnated Heart

While you get a new physical heart at some time after reincarnated-inception, your old memories from the heart, [loves, friendships, desires, oneness with God] come along with your original soul to make a new person. A reincarnated person has a new spirit, and body, everything else would be from a reincarnation soul and its associated Heart memory component like Ezekiel told us.

> ***Ezekiel 11:19*** - *And I will give them one heart, and I will put a new spirit within you; and I will take the stony heart out of their flesh, and will give them a heart of flesh:*

Reincarnation is odd to say the least. It is when a soul reenters a new body. Most generally, it is the body of a baby and this person is almost completely different than the original. His memories are different, his demeanor might be different, but there are "things" connected with the soul that make him the same. Rather than explaining, it might be better to simply list a few verses that can help here. Afterwards, I will try to put perspective on them. Throughout the New Testament, you are teased about John the Baptist being the reincarnated Elijah. The interesting part is the John [or Elijah] had no memory of being Elijah, but he still had the same feeling, faith, force, and function to continue his work. Whenever John/Elijah saw Jesus, grown up, he immediately worshiped him. You would think that John growing up with Jesus would naturally think Jesus was just a guy. Most of the people around Nazareth thought Jesus was a false prophet but John/Elijah had no issue because that part of our memory [love, etc.] is not lost. Those loved ones, still feel the same love and there can be a closeness from that memory.

Matthew Chapter 11 and 18- *"For all of the prophets and the law have prophesized until John. And if you are willing to receive it, He [Jesus] is Elijah who was to come." -- 'Why then do the scribes say that Elijah must come first?' But he answered them and said, 'Elijah indeed is to come and will restore all things. But I say to you that Elijah has come already, and they did not know him, but did to him whatever they wished. So also, shall the Son of Man suffer at their hand.' Then the disciples understood that he had spoken of John the Baptist."* This shows a strong belief in Reincarnation, so what happened to John the Baptist/Elijah?

Matthew 17:1-13"*After six days Jesus took with him Peter, James and John the brother of James, and led them up a high mountain by themselves. There he was transfigured before them. His face shone like the sun, and his clothes became as white as the light. Just then there appeared before them Moses <u>and Elijah</u>*- So, john, now Elijah came back to life, talking and completely visible, but then he vanished. <u>These guys had been dead for hundreds of years.</u> They had not been to heaven because Jesus told everyone that he was going to make heaven livable so you ask,

"Where were they?" Well the next section tells us a little bit concerning Elijah and where he has been.

***Isaiah 53**-After <u>the suffering of his soul</u>, he will see the light [or take in the Holy Spirit thing] and be satisfied; by his knowledge my righteous servant will justify many, and he will bear their iniquities.* [This is identifying everyone as being iniquitous, promising substantial suffering, presenting justification and understanding by God Incarnate, and finally this resurrection thing. Please notice that it is talking about the SOUL suffering rather than people suffering. Why do you suppose that is? I'll tell you. The Soul extends beyond one life.]

***Isaiah 66**-"From one New Moon [birth] to another and from one Sabbath to another, <u>all mankind will come and bow down before me, says the LORD</u>. And they will go out and look upon the dead bodies of those who rebelled against me; their worm will not die, nor will their fire be quenched, and they will be loathsome to all mankind."* [We don't know what else Isaiah might have written for his book ends with this final warning. The rebels mentioned here seem include all those who would not bring in the Holy Spirit. This included those who died well before the time of Jesus. The only way all mankind worship God and view the dead rebels is <u>for most people to take this Holy Spirit in during one of several successive lifetimes</u> allowing for the acceptance.]

Reincarnation or After Life Instruction

Daniel 11 and 12-*"Those who are wise <u>will instruct</u> many, though for a time they will fall by the sword or be burned or captured or plundered."* People will be teaching others even after they have fallen by the sword sounds like one can instruct after death. While there might be some way for a dead person to instruct, we can more easily understand how a person can be resurrected and pass on things he learned from previous lives. These would not be memories, they would be important things like understanding, humility, hope, love, charity, God, and empathy. He would instruct those things. *"Some of the wise will <u>stumble, so that they may be refined,</u> purified and made spotless until the time of the end, for it will still come at the appointed time."*

EVEN the wise will "stumble" in one life, but they will have more chances until the end of time. I think this verse is pretty clear. Reincarnation is to allow the most people to be saved for the final resurrection.

"Multitudes who sleep in the dust of the earth will awake: some to everlasting life, others to shame and everlasting contempt."

This last verse is the most important one. There will be only a short number of chances for reincarnation and then all is lost.

The verses say, after a number of reincarnations, there will be a resurrection. If, after many stumbles, one finally takes the Holy Spirit, this resurrection will be neat and while you are "living", your vibration level will be faster so you will be able to affect other people in positive ways during your life and AFTER.

Fight Against Entropy

Man cannot fight carnality without the aid of the Holy Spirit and that cannot be used without the Heart-brain communication channeling spiritual memories. After death, one would have the same issues. When the witch of Endor revived the "soul" of Samuel, it had "Spiritual memories" because of the Heart-brain memory transfers during life. Those without those "Spiritual memories" one. I don't know how all this happens, but there are reincarnations and there has been a long delay before the final resurrection for one thing as indicated in the following verses and others.

*2 Peter 3:8-9…But do not let this one fact escape your notice, beloved, that with the Lord one day is like a thousand years, and a thousand years like one day. The Lord is not slow about His promise, as some count slowness, but is patient toward you, **not wishing for any to perish but for all to come to repentance.***

God wants as many to accept the Holy Spirit as possible**.**

1 Timothy 2 :3-6 …For this is good and acceptable in the sight of God our Savior; Who will have all men to be saved, and to come unto the knowledge of the truth. For there is one God, and one mediator between God and men, the man Christ Jesus; Who gave himself a ransom for all, to be testified in due time.

God wants as many to accept the Holy Spirit as possible.

God's Return Is At Hand-If this reincarnation/reentry capability to increase our possibility of gaining the holy spirit was not so, the thousands of years God stays away does not make much sense. Jesus continued to tell his disciples that *the time of God's return was at hand*. They thought it was to be in their lifetime. In one way, it was as more and more believers are collected for his final return and we are not aware of any time spent between lives or in previous lives, so to us the time is still at hand.

Life-Certainly, there are many more commonly believed and exotic life and death ideas, but I really want to focus on increasing your vibration level and developing Self Actualization or whatever you want to call it so you can be aware of a truer reality where people can move mountains and walk across water and even be remembered by the dead. Let me say a VERY IMPORTANT CAUTION here.

Caution!!!-Anyone thinking he or she will get more time to live as debased as one can, will assuredly lower vibration level of our entire reality and they will be drawn away from the important part of life and the ability of possibly have communication with the departed. I am not saying they will not be able to gain the LIGHT [holy Spirit]. Rule of thumb is "without light we experience life as a dissipating element. When we get to its end, it somehow disappears. There will be NO REALITY for them. They will be in a non-life condition similar to that experienced by the Nephilim demons, but it seems that sometimes, God grants reincarnation for some. I don't know the mechanism or how often, but we are told that God has waited 2000 years for the final resurrection just to ensure that as few people as possible will not eventually gain the light and survive.

Long Term Memory

Long Term Memory of the Heart-brain-The ganglia in the human heart are interconnected, and consisting of many different types of neurons, like for example <u>local circuit neurons</u>. These neurons are known to be important in the hippocampus for encoding and decoding of long term memory.

It has been proven now that heart neurons also have long and short-term memory. The heart needs functional memory to do its job right, fast reactions are impossible if a signal has to be sent to the brain first for approval. Long term memory allows the Heart to remember even beyond our life.

Beyond Death

Here is possibly the most important part of all of this. When you die, the long-term memories of the Heart brain are not lost like those of the Head-brain. I'm not sure how all of this works, but I can tell you what I believe. The Bible indicates that when we die and potentially go to heaven, we will not be encumbered the horrors of the memories of loved ones being sent to hell or any of that and we are told by incarnate God, and by affirmation by John, that John-the-Baptist had no direct recollection of being Elijah in a past life even though Jesus himself explained over and over that John WAS Elijah. We are also told of small children that remember bits and pieces of a previous life or they speak an unknown language or know things they could not have known and these images fade at an age of about 8 years old. These pre-life memories are not self-directed but instead are event memories [Memories of capability, recognition, understanding, but not of self-adoration or sexual imaging], therefore, we can assume the transfer medium of memory to a next life is by means of the tiny, heart-brain long term memory and there isn't much of it. There are 2 places the memories could be stored; outside the body in this universe or outside the body in our linked universe [heaven]. If this is so, the memories would be distributed either by the Soul or spirit and recovered by the same methods.

The bible tells us that the instant we die, our spirit is transferred back to Heaven and that our entire dead body, including our head-brain turns to dust so the only memories left would be those that left with the soul that is more closely tied to the Heart-brain than we could imagine.

Ecclesiastes 9:5 *For the living know that they will die; But the dead know nothing [are asleep]*

Psalm 146:4-*His* spirit departs*, and he [the body] returns to his earth; In that very day, his plans perish.*

Ecclesiastes 12:7- *and the [body] dust returns to the ground it came from, and the* spirit returns to God *who gave it.*

Hebrews 4:12- *"For the* word of God *is quick, and powerful, and sharper than any two-edged sword, piercing even to the dividing asunder* of soul and spirit..."

John 3:5-6- *"When you are reborn, you are born <u>with a new spirit</u>"*

1 Thessalonians 5:23- *"And the very God of peace sanctify you wholly; and I pray God your whole <u>spirit and soul and body</u> be preserved blameless unto the coming of our Lord Jesus Christ."*

Mark 14:38 *"The spirit indeed is willing, but the flesh is weak"* [In this case, the Heart is having to service the "flesh" [self] so much that messaging from the spirit cannot be transferred through the heart.]

John 4:23 *"true worshipers will worship the Father in spirit and truth".*

1 Corinthians 15:20 *"But now Christ is risen from the dead, and has become the first-fruits of those who have fallen asleep."* [He awoke with a new Spirit and an old soul that had not gone to sleep.].

Luke 23:46: *Jesus called out with a loud voice, "Father, <u>into your hands I commit my spirit</u>." When he had said this, he breathed his last.* – [He immediately lost his Spirit, but his soul stayed nearby.].

Corinthians 15:3-*Christ died- and that he was buried, and that he <u>rose again the third day</u>* [With a new Spirit and an old soul].

However rebirth occurs, when reincarnation happens, some of those Heart-brain memories are "remembered". There are actually a lot of researchers in the field of reincarnated souls, but I want to concentration the Heart in this book. Towards the end of the book, I will review some of today's finding from other reincarnation studies for completeness but let's look at these memories as there is an accounting so some portion of our memories must be saved to allow for passed feelings, memories, loves, and friendships. Dr. Stevenson of the University of Virginia, has collected 1200 cases. Not believing past life regression was reliable and older recounts could be tainted by information collected from books or TV, he concentrated on asking children.

Hanan to Suzanne- *Hanan was a woman who died at a fairly young age and became Suzanne 10 days after her death. Born in Missouri, Suzanne remembered her old husband, all her children and family and described all 13 by name allowing researchers to find them for verification. Suzanne still loved her husband from her previous life and remembered their phone number. She began calling him when she was five and her parents allowed her to visit so she could sit in his lap and be with him from time to time. Even at age 25 Suzanne still called her old husband.*

Rashid to Daniel - *Born in 1969 Daniel's first word was Ibrahim, Rashid's close friend. When he was 2, he begged to go home, indicated his mom wasn't his real mom and his daddy was dead and he remembered his old father's mane "Naim". At 2 ½ years old he spelled out Kfarmatta, the town where Rashid was born. HE also remembered where he died, his old occupation, and how he died.*

Anne Frank to Barbo- *This has been Hollywoodized, but similar except Anne was in a concentration camp during WWII. Barbo remembered intricate details and provided enough details to her parents to allow them to find Anne's old home. A book was published when she was 10 years old and she became a writing prodigy just like Anne Frank had been.*

Carol Beckwith to Robert Snow- *This transfer was only found by regression hypnosis when Robert was older, but he told the therapist all types of intricate details about his previous life including details of paintings he had done. By chance he and his with saw one of the paintings he had done as Carol and he found that every single thing he had remembered was of Carol Beckwith's life.*

Xenoglossy

Xenoglossy is speaking a language you never learned in this life.

Grechen to Dolores- *In another of these hypnosis sessions Dolorese spoke fluent German and explained her past life in German 10 times and even wrote in German, all confirmed by the German interpreter used but the writing was not in her own handwriting. When awake Dolores still has no capability of speaking German.*

Jensen To Tania- *In this instance the previous life was a Swedish man who spoke Swedish instead of English of Tania. The previous man could also speak Norwegian. He was a farmer who was married to a woman named Latvia. He was killed by a Russian soldier. Tania didn't know or understand anything about German or Norwegian when awake.*

I've decided to not got through all 1200, but I think you get the picture. Almost as soon as a child can talk, many seen to remember vividly critical items of past lives and later they are forgotten, but sometimes can be recalled by hypnosis. These things include, names, phone numbers, language, loves, locations, intense pain. All somehow survive death and live "IN" the soul. Souls seem to enter a fetus even months after inception, so we cannot understand when people change from a life into a life with a soul and observer.

Heart-Brain Memories after Death

As the head-brain itself is destroyed as part of the death action. It embodies our carnal thoughts, images, action, memories and all the rest and none of that is very useful after carnal life leaves. The memories, feeling, loves, concerns, and stored events of the Heart brain appear to be the part of our memory kept alive. Some memories are transferred to our soul after passing, but some memories stay for some time and when our heart is separated from our bodies and used for transplanting, those feeling and memories are attached to the recipient's soul. As far as remembering elements of life after reincarnation, we have heard that some children speak a foreign language at an early age when they have no contact with that language and idiot savants that can play piano without learning how. Those are the types of memories carried in the Heart-brain for possible retrieval in a next life and I brought out earlier how transplant recipients have these same images. This is all possible because of the close union of the Soul and Heart-brain. The more we work to expand this link the better we will be.

Michael Newton's Soul Journey

Michael Newton is a hypnotherapist in California and he is good at it. His specialty is regression therapy. As many others have found, sometimes, fears or limitation we have in this life come from events that occurred in a previous life. Too many people have been regressed for it to not be considered a reasonable method of research and it cures people. Anyway!!! Michael started getting information from his patients that didn't make sense. They were talking about people they had contact with BETWEEN LIVES. Case study after case study, more and more data was collected and confirmed by the details presented by others. Soon, a reasonably clear picture of this purgatory type place you have heard about and dismissed. When events of a life are not completed in a reasonable way or the person feels anguish at something that happened or he did while alive, he relives those events on the other side to experience the other side and gain empathy. I'll bet you thought you got empathy because of your marvelous insight to your surroundings. Well!!! There is a growing amount of data that suggests that empathy comes from ACTUALLY living a life as the OTHER individuals. This "feeling" seems to be one of the strongest characteristics of what we have been calling the Heart-brain. Maslow's Self Actualization level that forces the insight concerning those around us, may not be a carnally learned event, but is accomplished over many existences.

Helpers-Here is the other thing Michael found. There are Helpers in this awaken state of "free souls" that help us understand what our previous weaknesses were. There are also un-embodied friends or acquaintances and groups that sort of stick together in a journey that possibly continues for a number of reincarnations. I don't know much about the helper souls, but maybe some don't reincarnate or they possibly have become "God followers" and wish to get others to follow by some type of reincarnation which goes along with the following verses.

***2 Peter 3:8-9**... The Lord is patient toward you, not wishing for any to perish but for all to come to repentance.*

__1 Timothy 2 :3-6__ God will have all men to be saved, and to come unto the knowledge of the truth.

More Reading-Michael's books, "Journey of Souls" and "Destiny of Souls" provide a very good overview of this new quai-science of intra-life regression so don't just take my word. Also, read "Many Masters" by Brian L. Weiss M.D. or contact a growing list of others around the world. Below are a number of Life between life therapists experimenting in this exciting new area where we can gain more understanding of death---while we are alive.

Australia- *Tony Collins, Tania Dionisio, Tracey Robins*

Czech Republic- *Bernadeta Hodkova*

France- *Kathy Gibbons and Bernadeta Hodkova*

***India*-** *Neeta Sharma, Blossom Furtado, Asis Ganguli, Jyotika Chhibber*

Ireland- *Kathy Gibbons, Izabela Fouere*

Norway- *Lise Stjernholm, Tone Hansen, Lisbeth Lyngaas, Kjus Garden, Helen Soernes, Bodil Rosvik, Alice Kjolerbakken*

Singapore*-Reena Kumarasingham, Peter Mack, Antoinette Biehlmeier*

Sweden*-Lis Lindahl, Karin Danneker*

Switzerland- *Radovanovic Kchler, Jacqueline Niggli*

UK -*Hazel Newton, Liz Kozlowski, Elen Clulow, Reena Kumarasingham, Ian Lawton, Peter Blayney, Maggie Amuna, Janet Treloar, Anjalee Carey, Lorraine Flaherty, Wissam Awad, Reena Kumarasingham, Doug Buckingham, Tricia Allen, Bridget Rattigan, Liz Kozlowski, Katherine Membery, John Nicol, Chris Hanson, Dave Graham, Debbie Wild, Linda Hopkins, Janet Treloar, Trish Heenan*

It looks like if you visit England, you can find out a lot about what this book is about. I wonder how many hypnotherapists DON'T try to determine what your Heart-brain memories and souls did between lives??

That little tirade on reincarnation and what that has to do with the heart-brain, let's switch over to another seemingly strange subject of circumcising your heart.

Circumcised Heart

I know every time someone says the word circumcision, you think about a small piece of penis being cut off as was done by the ancient Egyptians and by the Jews under direction of God's command. The following drawing shows how the Egyptians used to do this act first cutting with a scraper of some type while the victim said things that were only expressed as groans of something and then a bandage was wrapped around. Egyptians were proud of their new look and would go around naked to show off, but Jews were told something different.

They were told to separate themselves from the rest of the world and the cutting the penis thing was just a symbol. It would have been between the Jew and his god. OK! The Jews never really understood the whole circumcision thing and did not separate themselves from the Canaanites or anyone else, they just had a shorter penis. To make the point, the prophets started describing circumcision of the ear, eye, tongue and heart. Of these the most important Circumcision was of the Heart-brain. God wanted the Jews to be circumspect in all that they did that might go against God's law. While the penis controlled going against God's law of intermarriage and fathering half-breed children, the Heart-brain needed to be focused on communications with the Spirit as the spirit could communicated with God.

> ***Jeremiah 9:26*** *even the whole house of Israel is <u>uncircumcised in heart</u>."*

The soul communicates outside the body, but it is still restricted to our own universe. Circumcision of the Heart was the main thing needed for salvation and many Jews were using the laws of sacrifice to continually be forgiven of egregious wrongs initiated and conducted by having and uncircumcised heart. As an aside, the penis cutting circumcision practice continued in Medieval time and there still was not the slighted understanding of why someone would need a shorter penis. The image following shows this continued practice.

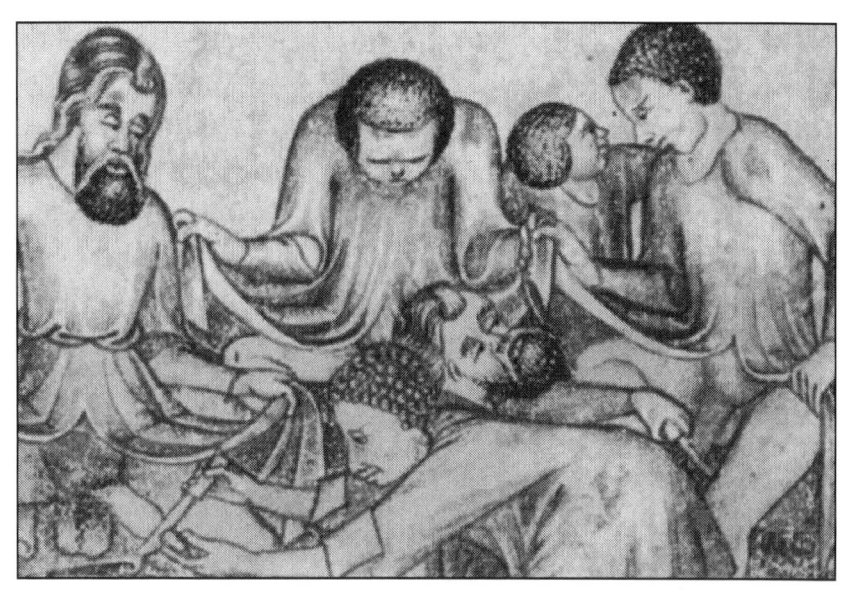

Genital Mutilation

While we are on this subject, I need to discuss this so called female circumcision, clitorectomy or infibulation. Certainly snippig off a piece of useless skin on a penis is not mutilation, but what Muslims do to women is simply horrible and somehow tied to staying reverent to tghie Moon God, Allet or Allah. One of the highly vocal women against this barbaric practice said she was first cut when she was five, then again when she was 16 and a third time when she was in her twenties. While any of is is totally wrong, there are different levels of this horror. They include removal of the clitoral hood and clitoral glans. A second method is the removal of the inner labia, but the last method is almost too hard to write about. It is the removal of both the inner and outer labia and closure of the vulva. This is called infibulation and only a small hole is left for the passage of urine and menstrual fluid. After infibulation, the vagina must be reopened for intercourse after marriage and opened more for childbirth. Now for the kicker; this barbarism is still going on and over 200 million women and girls in 27 African countries, Indonesia, Iraqi, Kurdistan, and Yemen, have endured it as determined in 2016. I'm sorry for the last part, but I am sick of people trying to make circumcision of the genitals as something holy and just.

What About Circumcising the Heart?

Jeremiah 9:26 *even the whole house of Israel is <u>uncircumcised in heart</u>."*

Jeremiah 4:4 *"<u>Circumcise yourselves to the LORD And remove the foreskins of your heart</u>, or else My wrath will go forth like fire-because of the evil of your deeds."*

Romans 2:29 *<u>circumcision is circumcision of the heart, by the Spirit</u>, not by the written code.*

Romans 3:30 *-since indeed God who will justify the circumcised by faith and the uncircumcised through faith is one.*

Philippians 3:2-3 *-Beware of the false circumcision; for we are the true circumcision, who worship in the Spirit of God and glory in Christ Jesus and put no confidence in the flesh*

Colossians 2:11-12 *-In Him you were also circumcised with a circumcision made without hands, in the removal of the body of the flesh by the circumcision of Christ.*

Deuteronomy 10:15-Yet on your fathers did the LORD set His affection to love them-- So <u>circumcise your heart, and stiffen your neck no longer</u>.

Deuteronomy 30:6- Moreover the LORD your God will <u>circumcise your heart</u> and the heart of your descendants, to *love the LORD your God with all your heart and with all your soul, so that you may live.*

Spirit Focused Heart-Brain

In the Sermon on the Mount, recorded in Matthew 5:8, Jesus specified, *"Blessed are the pure in heart, for they shall see God."* Purity of heart, described by the Orthodox Church in America, is freedom from wicked motivations and evil intentions, refraining from self-seeking desires, and freedom from anything that clouds the mind from seeing things clearly and honestly. In the Orthodox tradition, purity of heart-brain is essential for unity with God or unity with God allows for purity of the Heart-Brain. Only an active heart-brain separated from the Head-brain could this be possible as it requires the direct communication with the spirit and adherence to the wishes of the spirit that communicates with God. There is no such thing as pure in Head-brain as we are told over and over that man thinks only evil continuously and it happens to be true. As selfish pride is one of the worst things one can do, how many people have more than one mirror to make sure their hair is just in the right place?

This is Faith #1

The Heart-brain becomes pure from direct and open communication with the Spirit that communicates with God. One must allow the Heart-brain to accept the words, feeling, issues, concerns, and love from Heaven to attain this purity. With a pure heart, salvation can be readily attained. Once salvation occurs, the bond between the Spirit and the Heart-brain would become stronger and stronger applying interest and desire for us to help others obtain this level of unity. The Sermon on the Mount said it best and a pure heart-brain will allow the host soul to see God. There is little doubt that the Bible is trying to get you to recognize your heart-brain and use it to obtain purity through the spirit of God. The only thing left for repentance is acceptance of God as the savior and following the Heart-brain descriptions given by the Spirit of God.

Genesis 24:45 *"Before I [servant of Abraham] finished praying in my heart--- "*

1 Samuel 2:1 *My [Hanna] heart rejoices in the LORD;*

Jeremiah 20:9 *His word is in my heart like a fire,*

Jeremiah 20:12 *you [God] who examine the righteous and probe the heart and mind,*

Psalms 51:10 *- Create in* me a clean heart, O God; and renew a right spirit within me

Psalm 77:6 *My heart meditated and my spirit asked:*

Psalm 90:12 *we may gain a heart of wisdom.*

Proverbs 3:5-6 *- Trust in the LORD with all thine heart; and lean not unto thine own understanding.*

2 Chronicles 30:19 *who sets their heart on seeking God*

Nehemiah 9:8 *You found his heart faithful to you.*

Mark 12:33 *To love him with all your heart, with all your understanding and with all your strength,*

Hebrews 3:12 *none of you has a sinful, unbelieving heart that turns away from the living God.*

Philemon 1:20 *<u>refresh my heart</u> in Christ*

As you might expect, researchers have not made up tests to observe this capability of the Heart-brain; however, the graphic below shows the focused characterization of the heart-brain towards the Spirit entity of our being that is needed for purity, repentance and the gain of spiritual wisdom.

Heart-Brain Repentance

Repentance of the heart was an essential teaching in the Bible. In *Psalm 51:10*, David asked God for a clean heart and a right spirit. The Hebrew words "niham" and "shub" in the Old Testament refer to repentance. According to the Jewish Encyclopedia, niham means to "feel sorrow" and shub means "to return." Saadiah, a Jewish philosopher in medieval times, felt that true repentance requires regret and remorse for the sin <u>by the heart</u>, as well as renunciation, confession, asking forgiveness and pledging to not repeat the sin which was done by the Head-brain.

One cannot be repentant until the Heart is repentant. We can determine that the communication between the soul, spirit and body is regulated through the Heart-brain.

Testing the Heart

Several of the texts talk about God testing the heart to see if we were in line with his teachings. With the heart brain being an active component of life, God would need only to see how well the spirit communicated and was used by the heart brain. If the heart rejected the spirit inputs, there were serious problems. Unfortunately, many try to limit the actions of the heart and run into trouble every time.

The **Holman Bible Dictionary <u>incorrectly</u> explains** that the Hebrew language had no word for conscience, so the word heart was used in place of conscience to explain the concept of deciphering right from wrong. Besides the obvious issue with the Greek having a word for conscience and, still, Heart is used; texts don't seem to fit the Holman definition.

Jeremiah 17:10 *the Lord searches the heart and tests the mind to judge every person by the fruit of their deeds.* <u>Conscience doesn't fit.</u>

Deuteronomy 8:5 <u>*Know then in your heart*</u> *that as a man disciplines his son, so the LORD your God disciplines you.* <u>Conscience doesn't fit.</u>

Deuteronomy 8:14 *then your* <u>*heart will become proud*</u> *and you will forget the LORD.* <u>Conscience doesn't fit.</u>

Hebrews 3:12 *none of you has* <u>*a sinful, unbelieving heart*</u> *that turns away from the living God.* <u>Conscience doesn't fit.</u>

Hebrews 4:12 *the word of God penetrates even to dividing soul and spirit; it judges the* <u>*thoughts and attitudes of the heart.*</u>

Hebrews 10:22 *draw near to God with a* <u>*sincere heart*</u> *having our* <u>*hearts sprinkled to cleanse us from a guilty conscience.*</u> <u>Conscience doesn't fit.</u>

Various Christian doctrines preach that God allows people to willfully choose obedience or disobedience. In a biblical context, God looked at the heart to see where people stood in their faith and deeds. Again, Conscience doesn't fit.

The reason this description is wrong is those trying to push the definition did not even think that the heart could have an active role in our lives so let's continue by looking at what the Bible called purity of the Heart.

Spirit Wisdom

Why in the world would Solomon think the heart was wise when he was the wisest man in the world? The answer is he knew wisdom [as defined in the Bible] came from interactions between the spirit and the Heart. They were not initiated by the Heart, but they are received by the Heart-brain. Without the Heart-brain, a link with the Spirit is precarious. The book of Isaiah provided the quick list of interactive qualities established by the Heart-brain interface with the "Spirit".

***Isaiah 11:2** The Spirit of the LORD will rest on him—the Spirit of <u>wisdom</u> and of <u>understanding</u>, the Spirit of <u>counsel</u> and of <u>might</u>, the Spirit of the <u>knowledge</u> and <u>fear of the LORD</u>—*

As we see, while the Heart-brain mostly works on our being, wisdom is not specifically "from the Heart". It comes from the third part of our three-part being [Soul Self, Spirit] and the heart is the intercessor between the spirit and the soul or self. They are described next so you can see the important differences in these characterization in the Bible.

Wisdom: *It is the <u>capacity to love spiritual things more than material ones</u>; and the want to understand God. It is what allows for redemption.*

Understanding: *Like the first one, this is spiritual understanding. A person with understanding would not be* *confused by the conflicting messages in our culture about the right way to live. It is the gift whereby self-evident God sent principles are known.*

Right judgment: *With the gift of right judgment, we <u>know the difference between right and wrong</u>, and we choose to do what is right. A person with right judgment would avoid sin, for instance.*

Courage: *With the gift of courage, we overcome fear needed to follow the Incarnate, Creator God. This allows a person to do good and endure evil.*

Knowledge: *With this gift, we understand the meaning of God and helps us to choose the right path through life;*

Reverence: *With this gift, we have a deep sense of respect for God. A person with reverence recognizes <u>our total reliance on God and comes before God with humility</u>, trust, and love.*

Fear of the Lord: *With this gift, one KNOWS that God is the perfection of all we desire: perfect knowledge, perfect goodness, perfect power, and perfect love.*

We are told without the Spirit link to God these things cannot be obtained and direct communication with God in our linked universe is impossible. In the book of *"Romans"* we find we can't even pray without this <u>Spirit talking through our Heart-brain</u>. Here is what it says.

Romans 8:26-27-*Likewise the Spirit also helps us in our limitations for we do not know what we should pray for as we ought: but the Spirit itself maketh intercession for us with groanings which cannot be uttered. And <u>he that searcheth the **hearts** knoweth what is the mind of the Spirit,</u>*

because he maketh intercession for followers according to the will of God.

There you have it. The Spirit entity is a go-between to provide all of the things I mentioned and change what you pray for to what is respectable and responsible for the Creator of the Universe to answer and then the spirit interprets that into something the Heart-brain can send to the head and soul. I know it sounds complicated, but talking to the creator of the universe is a pretty big thing. By the way, most prayers you hear people say aloud are not getting to the spirit. They are made for people to hear and are of no specific use to the Heart-brain.

The following graphic describes how a spirit or God centered Heart-brain allows for establishment of wisdom. As self is thought about less and less, the Soul and Spirit begin increase in significance and power all channeled by the Heart.

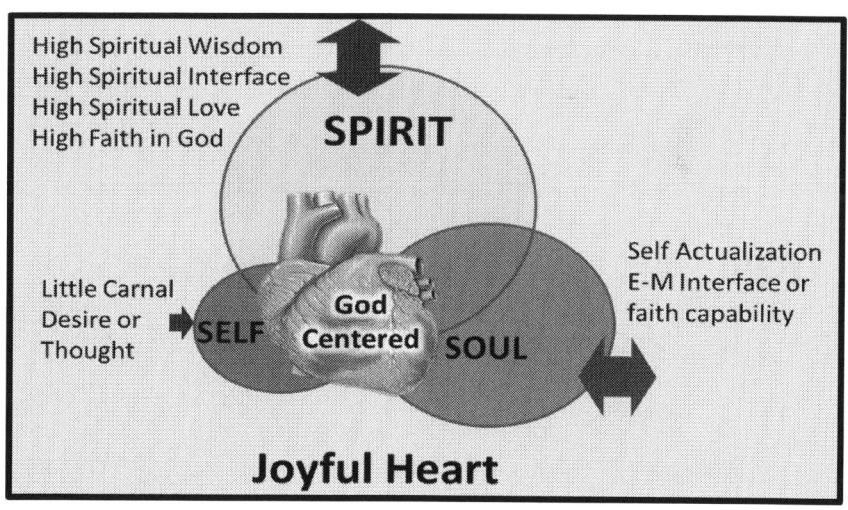

***Matthew 5:8** Blessed are the <u>pure in **heart**</u>, for they will see <u>God.</u>*

Romans 10:9 *believe in your **heart** that God raised him from the dead, you will be saved.*

1 Timothy 1:5 *love, which comes from a pure **heart** and a good conscience and a sincere faith [#1].*

Deuteronomy 4:29 *seek the LORD your God with all your **heart** and with all your soul.*

1 Samuel 2:1 *My [Hanna] **heart** rejoices in the LORD*

1 Samuel 12:20 *Samuel replied, serve the LORD with all your **heart**.*

***1 Kings 3:6**—he [King David] was righteous and upright in **heart**.*

Psalms 51:10 *- Create in me a clean **heart**, O God; and renew a right spirit within me*

2 Chronicles 30:19 *who sets their **heart** on seeking God—*

Nehemiah 9:8 *You found his **heart** faithful to you [God].*

Please notice the Heart-brain is required for all of this to happen and it is addressed in most of the verses specifically. Open you heart to understanding the spiritual characteristic of existence----the creator God and all that he has done for his creation. Without listening to our heart concerning these things with be devastating after this temporary flight though this life. You can get guidance in expanding the awareness of the Soul and interface with our true reality by descriptions and meditations for what are called chakras as the soul is not a spiritual entity so we will look at these descriptions. That being said, the crown chakra is described, but not attainable without invigorating the spirit portion of your entity.

Chakra & Heart Brain

As I have described, we know the Heart changes the cadence of the brain by E-M emissions. One method used by some religions is to meditate on elimination of self by enhancing a charka. Those practicing these deep meditations indicate that their whole body seem to vibrate and the faster it vibrates the more "open" one becomes. Certainly, the first step is rejecting the body. The Taoists said it the best.

> *"Our body should be an empty vessel."*

The Buddhists and other groups call the various levels of vibration and awareness, Chakras; which is just an indicator about how you feel and acknowledge this "reality". The reason I brought up this effect is that it describes how sensitive our consciousness is to vibration. The reason is simple. Consciousness is a vibrational dimension and brainwave studies are not the only way to recognize the vibrational characteristics. A second way is something called "chakra" by Buddhists so let's look at some of these mystical things. According to the believers and the testing skeptics there are at least seven of these chakras or levels of consciousness. As someone increases his chakra, we are told he feels a vibration all around and inside.

They are sort of represented by the listing following.

- Root, Chakra--- "Consciousness of Survival"
- Sacral Chakra --- "Consciousness of Sex"
- Solar Plexus Chakra --- "Consciousness of Self"
- Heart Chakra --- "Consciousness of love"
- Throat, Chakra --- "Consciousness of the truth"
- Third Eye Chakra --"Consciousness of inner being"
- Crown, Chakra ---"Consciousness of the spirit world"

As a pictoral description, Chakras are shown to start at the root of the body and work higher and higher until you compley leave you body and this reality in what is called the Crown Chakra.

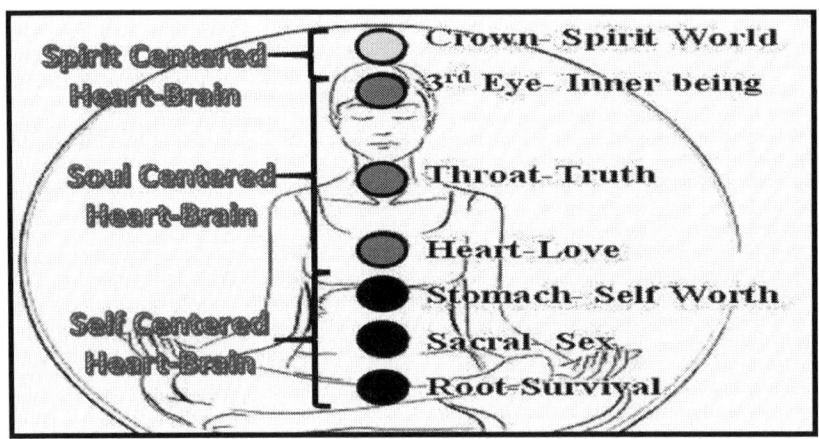

I know it sounds like I'm some guru from India talking about chakras, but it is a convenient way to discuss the heart-brain in a different light so I'm going to continue. I'm not putting on the towel on my head, but I may hum a little as I write this section.

Remember what this is all about is the rejection of the Self, Survival, Sex debased thoughts that restrict knowledge, wisdom, and enhanced control over our environment to essential do what the Bible recommend.

If you want to live you must die [so to speak]. This does not mean kill your heart-brain---just everything else that attempts to force you into a "carnal existence. The Heart-brain does the opposite.

Let's say a girl witnesses her father being pinned under a car. She immediately rejects self, sex, and survival and this allows her to "move mountains"--- or a car in this situation. As soon as she realizes what she did, reality goes back to normal.

Elisha walked on water by doing the same thing as his body no longer is heavy enough to sink.

John Hutchison's experiment making a bowling ball rise off the table is the same, but he simple introduce ultra-high beat frequencies on the scene with his equipment to momentarily change reality in the "focused" area. I'm not saying his method ill help the Heart-brain heal the sick, but it shows that our "Reality" is based more on what we "Think" than what is.

Bowling Ball Flinging

While some of the things the Canadian Researcher, John Hutchison, has accomplished in the last few years are remarkable, let me just show you a couple of the frames from one experiment in the presence of a powerful and strange field of electromagnetic waves that could be a general analogy to the Heart-brain Electromagnetic waves. First pliers were picked up and yanked out of the view, but then, a bowling ball rises off the table until the ultrahigh frequency E-M vibrations were halted and the bowling ball resumed its normal heaviness in our reality.

I must say that John typically has trouble duplicating his results as the actual beat frequencies are very difficult to record, but they are still something special. With that as an introduction, let me get into more about Participatory Anthropics as it shows that the Heart-brain and even our head-brain could modify reality or heal the sick.

Soul Enhancement Levels

Survival, Self, and Sex

Whether we admit it or not, every one of us battles survival, self, and sex every day. The most basic root chakra is triggered when you get hungry, fear brings out survival, and the sex one, well; it shows up from time to time. I'm sure you recognize that these components of our life are, pretty much, uncontrollable and they should never be associated with conscious control. In fact, these "feelings" run parallel to conscious thought always ready to break or modify our Head-brain consciousness. I only brought them up again because some identify them as consciousness levels. A rule of thumb might be if it is associated directly with pleasure, pain, desire, or self-preservation, it is not part of the soul dimension of your life. Some self-preservation elements such as moving your hand away from a burning flame is identified as autonomous. I would contend that other things are autonomous as well.

Love and Truth

If you can get past those you start considering love. Most people spend most of the time going back and forth between the lower 3 chakra levels and the others. It's sort of like a yo-yo or they interact simultaneously. Slow vibrations [survival], higher vibrations [love], slower vibrations[sex], higher vibrations [Truth] and still higher vibrations when someone pays me a compliment or a figure out why a light bulb turns on in the refrigerator or I answer one of those "Are You Smarter Than a 5th Grader" questions. Answering one of the "Jeopardy" questions correctly might even get you into the love chakra or plunge you down into the Self-centered Chakra.

The love [or Heart] chakra is a vibrational level associated with "Real" love rather than the sexual one. I believe that an ameba and a tree have no capability of love, but they are still alive. Sometimes this "heart" level happens naturally for a brief time and you can't seem to even think about yourself at all. If you work at it you can get to this level periodically throughout a day and look at people with true look. The Bible called it loving people as you love yourself. Anyway, most just think they get into love and it is more basic. That type of love puts you below the stomach again. Anyway, you must conquer love to some level before you can even get to a point that looks for "real truth". Real truth is a truth that is truth no matter how it affects the event or who thinks it. It is usually not a popular truth or even the one you would hope for. It simply is. While it seems that this would be easy to understand and use. People almost never are tuned to this type of consciousness so they accept what they believe rather than what they should believe. Let's say you get an openness to understand real truth, the heart brain is getting more and more focused on the soul as the Head-brain needs less and less attention. The next chakra is called the third eye and this is the one that allows miracles to be performed.

The Third Eye

The third eye is derived from a little gland in the brain called the pineal "pinecone" gland. The pineal has no apparent use, but it is thought to have been used by our head-brains at one time. After all, the gland didn't just grow there for no reason, so let's look at what this thing is and how it might have been used by the Heart-brain. The following images show where the Pineal is in our head.

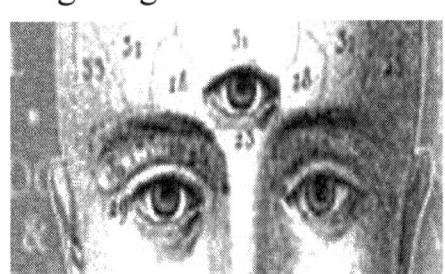 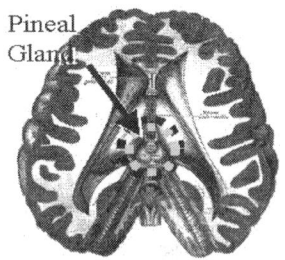

Tower of Babel

I could bring up many other things that would make you wonder if the pineal gland used to allow us to do many things in the past, but I won't. Instead, let me tell you what this tiny, pea-shaped gland does.

Pineal Glands in many non-mammalian vertebrates have a strong resemblance to the photoreceptor cells of the eye. Some evolutionary biologists believe that the pineal cells share common characteristics to retina cells in the eye.

In some animals, exposure to light of this gland can change the animal's biorhythm. Biorhythm is sort of code for the Heart-brain coherence level. If your biorhythms have a slow repetitive rhythm and amplitude, our become calm and relaxed.

Some early vertebrate fossil skulls have a pineal opening so that it probably had some vision characteristic.

The lamprey and the tuatara both have this same type of pineal opening and this thing is photosensitive. The structures appear to include cornea, lens and retina.

The pineal gland is weird in that it has profuse blood flow, second only to the kidney, so we can be sure that it once was of great importance. While doctors are perplexed at why this insignificant gland would need so much blood, it is obvious that whatever happened 6 thousand years ago made the extra blood flow unnecessary.

The brain of a 90-million-year-old bird was found with a large parietal eye and pineal gland so it's been used for some time now to provide additional insight beyond normal seeing.

Production of melatonin by the pineal gland is stimulated by darkness and inhibited by light. This melatonin stuff affects sex drive.

I hope you are seeing that this pineal appears to be an extension of the Heart-brain rather than an extension of the Head-brain. It is associated with feelings and insight. A long time ago, this gland in the center of our brain, may have been used to allow us to do all types of things we can no longer do.

Anyway, this chakra level is indicated as that needed to gain insight about the universe outside our "reality". This can only mean an even stronger communication link between the Heart-brain and soul and it may not have anything to do with the pineal gland, but it does have something to do with Abraham Maslow.

Abraham Maslow

This pineal gland/third eye was supposed to have given us the ability to understand the world around us. If we increase our vibrational level by unison with our environment "some call it meditation" or by other exotic means, we can sometimes get in tune with the world around us and here is the odd part. We can even affect it. Another way of saying this is that the 3rd eye thing is "Self–Actualization" that Abraham Maslow talked about.

Positive Thinking

Somehow getting our vibrational levels in tune with the vibrational patterns of the elements around us allows us to be more intuitive. We can sense reactions needed to affect the environment. As we affect the environment we can change it. Now the changes are extremely subtle. You cannot, for instance cause money to fly off a tree, but you can somehow affect the conditions around you that will make it easier to accomplish particular tasks simply by concentrating on these tasks and believing that these things will be accomplished. I know it sounds like gobbly-gook. The problem is that the affect is demonstrated over and over and over again. Positive thinking and getting in tune with the vibrational pattern of the environment actually works. There is no doubt about it and all you need to do is allow the Heart-brain to do its work and not be bogged down trying to keep you from thinking about your selfish needs. The issue is trying to get into the level of consciousness needed to get the universe to "Bend" a little is not only hard, it also is not easily sustained once one gets to this level of consciousness.

Anthropics

This brings us to a seemingly stupid question. *"If a tree falls in the woods, does it make a sound?"* The answer, according to Einstein and many others, is that it not only does not make a sound----- it simply does not exist.

The universe is "brought to life" by the characteristic dimensions of life [self, spirit, soul]

What is Matter

First, we need to know what matter is. Today it is defined as **Quantum fluctuations**. This is code for something Einstein called vibrating Aether with Aether being the *"potential for having matter appear"* as soon as this nothingness is seen by a "cognizant observer [person] it becomes real. As soon as they are not witnessing quantum fluctuations, there is no matter. The faster this "nothing" vibrates, the larger the particles as perceived. Gold, for instance is Quantum fluctuations vibrating at 600 Exahertz while sand/silicon is the same stuff vibrating at 8.5 Exahertz [8,500,000 Giga-Hertz]. By modern physics, to turn sand into gold simply shake it really fast. With healing, one must only modify the vibration of a cancer very slightly to make it become "normal" flesh. To walk on water, vibrate the water a little faster and it would become solid. To lift a car off someone, simply make the car vibrate slightly slower and it becomes lighter.

We See This Every Day

To see this happen we need only look at the stars; using Dr. Hubble's theory called red shift. Stars emit helium which makes a star look yellow, but some stars are moving away from us very quickly. Astronomers know where these stars are and they are very easy to identify as they become red because, to us the stars are now emitting lithium which looks red. Here is the thing! The yellow star and the red star are exactly the same except the fast stars all changed into lithium stars in OUR reality--- for our cognizant viewers.

Scientists tell us, without life, everything would have no REFERENCE so they would be nothing at all. We can take that a step farther and say without our heart-brain communicating with our soul that allows us to experience things outside our bodies, we would have no reference to reality.

People, *to be more precise—souls*, combine together to "create" what we can call the carnal world. This is the world we generate in our subconscious to be enjoyed by our consciousness <u>by communicating through the heart</u>. Without the heart communication, reality would be solely attributed by the self and true interface with others is impossible. One becomes aloof and finally sociopathic. It that state, a person can operate in an environment, but cannot sense other's feelings, needs, desires, requirements or anything else that makes us "human or humane". The theory of Anthropics allows the combined "resonances of the soul" [Heart-brain coherence between the outside world and the head-brain] portion of life to makes us understand what we call reality outside our selfish needs. Before we can understand how vibrations turn into a door, or friction, or photons, we need to understand the what is called the dimensional dynamo of life [self, soul, spirit]. All three are needed for life control of the environment. To be fair, reality in not just life, but also it is made up of 2 other things. Matter and Forces. Life, mass, and force are needed to define anything in the universe and the Heart-brain allows us to interact with this universe.

Let me tell you a definition you possibly have already heard.

Dr. Schrodinger had a cat that he placed in a box such that the cat might take poison and die or not. [Of course, he didn't really do this] He stated that the cat was both alive and dead until the box was opened. Once an observer reconciled "A truth" reality would establish the new course.

What he was saying just like Einstein and his tree falling in the woods disappearance, was that reality only exists now because cognizant observers make it so. Previous and future existence has no TRUE reality. What quantum fluctuations are all about is that even the reality we experience is very volatile as mass, and all forces are not completely real. Some would suggest the Head-brains establishes reality, but the more reasonable way to present reality is by the Heart-brain communication and link with our souls who somehow setup a reality acceptable to all enjoying it.

Let me give you another piece of insight.

In the book of "Revelation" all the souls are sent to either hell or heaven and then the book says something very strange. Remember Jesus had been making heaven a nice place over the past 2000 years or so, but all of a sudden something happened to the beautiful heaven in chapter 21.

Revelation 21:*1 Then I saw a new heaven and a new earth, for the first heaven and earth had passed away, and the sea was no more.*

Earth disappeared and a new on was required. There were no cognizant observers and reality stopped for an instant. The loss of the mirrored land in heaven has to do with the Physics of Super symmetry, so don't worry about that as we can determine that the heart-linked souls of this universe apparently ARE this universe. The part that the heart-brain memories play on the collection of souls holding our reality together may be beyond our understanding, but please don't discount the soul memories that shape our existence or you may end up like Schrodinger's cat.

More on Soul Control

Let me continue with this heart-brain infused collective of souls just a minute. The meaning of life is not 42 as indicated in a popular science fiction film. Instead, it is a complex variant of our universal existence. As I said before, if a tree falls in the woods and no one was around, the incidence simply does not happen. Without a cognizant observe, the universe modifications cannot take place and without time there is nothing that can be observed in our reality. So here is the problem. The soul is not directly attached to TIME. A thousand years is the same as a day. The Heart brain attaches time to our soul so that it can be used to affect reality. Once that happens, if the soul changes water into wine, it becomes real. That is not to say the soul is time because cognition seems to be "timeless" or accomplished adjacent to time. Because the soul has this freedom, people have been able to sense the future and see the past. People have been able to lift automobiles off people who were trapped without crushing their own bones. People have been able to walk on water and turn sticks into snakes and all the rest identified in our most holy historical references.

Now I will ask. If you see 700nm photons [called red light] and someone else sees 700nm photons do you see the same thing? Common sense would suggest that how the brain processes the various wavelengths of electromagnetic pulses would have to be different, but <u>if consciousnesses are linked</u> by means of the Heart-brain, when I see red, others see the same thing.

We are all attuned to this changing reality and we are responsible, in a small way for changing our perception of reality and the color red because of the Heart-soul communications of our being.

It doesn't matter how fast we go or how strange someone is, Red is not just defined, and it is established by our collective consciousnesses through our heart-brain communications. The more we communicate with the soul, the more control we have with reality [the more miracles we can produce.] We are finding the effect may not just be localized.

Heart-Brain Globalization

While we are on this subject, let's look at some of the newer findings of the HeartMath team concerning the heart-brain effect on the world and how music may greatly modify E-M signatures of those listening.

Globalization of E-M fields

The HeartMath global coherence study is trying to test how the world environment changes by localized heart-brain E-M communications. To do this study they are testing the Earth's magnetic field resonances and seeing how discord modifies same knowing that discord is transmitted by the Heart-Brian as variations in E-M frequency and variability and how modification may reduce instability and discord. One reason for this study is that the ionosphere around our planet, has a resonant frequency, called the Schuman Resonance, of about 7.8 Hz, that is very similar to the nominal frequency of human brainwave. Here are some of the topics of interest.

- Rhythms in the Earth's Magnetic Field seem to be associated with: changes in brain and nervous system activity, performance of sports

- Synthesis of micronutrients in plants and algae seem to be affected
- Mortality from heart attacks and vascular disease seem to be associated
- Depression and suicides seem to have a connection
- Number of strokes seem to be associated with instability of a region.
- E-M communication seem to affect the Earth's natural Rhythm.
- Geomagnetic Field right at the time of the attack on September 11th seemed to be slightly different than before or after.

It seems that E-M fields from a coherent heart goes out like a radio wave and modifies the resonance of the earth's geomagnetic field slightly to create more coherence in the environment. As we become more coherent ourselves, we can create a higher level of social coherence which may lead the world to be globally coherent which in turn will reduce stress, instability and discord. At least, this is the hope. On a closer view, HeartMath is studying music.

Music and the Heart-Brain

In 2004 HeartMath institute decided to see if coherence could be increased by listening to music and guess what! It can reduce blood pressure, and other variables associated with establishing coherence. After testing a number of music types, they found the following:

- Indian raga music was the most calming
- Rap the most arousing.
- Highly emotional arias, particularly by Verdi, used a 10-sec rhythm which affected the Heart-brain coherence the most.

Music modifies Heart-brain and head-brain communication and has been found to do the following:

- reduce stress
- improve athletic performance
- improve motor function after stroke or Parkinson's disease
- Possibly improve milk production in cattle

Expanded Review

Well I hope you have at least gotten something out of this book that will help you in this very strange life. As a review that might help jostle you brain, here are some of the major elements I tried to get across.

Heart- The heart is not just a pump. It controls the major hormones of the body, calms the head-brain, causes anxiety, love, and compassion. It can communicate with nearby people, cure the sick, and invigorate the soul to do all types of miraculous things. After a transplant, some of its feelings and memories are transferred, but the old soul is released allowing different interpretations for some memories.

Heart Brain- While we still don't understand the methods, we now know the strange discussions of the heart brain in the New Testament were really trying to tell us about this important part of our thinking that may last beyond death.

Bible and Heart- In 513 descriptive passages, the Bible describes in intricate detail the capability and control of the Heart-brain. It suggests the heart controls our lives. And some of these writing are almost 4 thousand years old. Not only is it in control of much of our lives, it is the gateway between our souls, spirit, and self and assures all are eq1ually addressed.

Hardened Heart- We find that the heart-brain can be hardened or focused on the Self portion of life if we do not allow it to extend its communications. While the "self" can survive with the heart hamstrung in this way, our lives are greatly limited and much less rewarding. This is accomplished by focusing more and more of our thoughts on self instead of having empathy concerning those around us. The Bible described it very simply. It said Love your neighbor like to love yourself or die.

Electromagnetic vibrations- Are controlling elements of life and a communication medium. Like radio signals, our heart brain manipulates blood pressure and hormone levels just like eyes, hands, and a voice. In some ways, it can see better than the Head Brains [eyes] because it can sense the truth in a person and his compassion.

Power of the Heart- Unlike the head-brain, the Heart-brain is not bogged down in housekeeping of Voice, sight, touch, hearing, walking, flexing muscles. and other things that take substantial focus and effort. Therefore, it can provide for long distance E-M communication of fairly simple messaging of feeling, love, concern,

Laying on of Hands- to strengthen Heart-brain communication, one can touch the intended recipient

Simulated Heart Brain machines- Today we have Infratonic Stimulators and Transcranial magnetic stimulation machines and we are finding music with just the right cadence, can do some of the things the Heart Brain does and can do.

Faith- We discussed the 2 types of faith while the Faith in God's redemption may be the most important to us after death, the Bible tells us to use our heart-brain to strengthen our interface without soul that can modify what we perceive to be reality. Walking on water, healing the sick, pulling demons out of people and turning water to wine were not side-show magic, they were simply the use of our Heart-brain being focused properly.

Nothing has an End-The important thing that many now profess including Einstein and Dr. Wolff is that matter never ends and life cannot end. Certainly, the body dies and turns to dust, but the major part of us, our soul lives on holding reality together and potentially being revitalized in a new "self".

Memories After Death- Our brain disintegrates, but we know some memories linger after death. It is believed these memories are initiated in the Heart brain and even stored in long term memory in the Heart, but also are transferred to the soul to be requalified in a following existence. [resurrected or reincarnated]

Life is three dimensional- like matter and energy, Life has three characteristics Life is made up of self, soul, and spirit. It certainly is not DNA, but most of the ancient religious data tells us that a being is made up of three entities [consciousness/self, soul/heart, and spirit/ connective communication]. They were called [Id, ego, superego] by Freud and [Ba, Ka, Shut] by the ancient Egyptians, but the important thing is the self is not the part we should be dwelling on as it can easily be swayed towards evil against reality or evil against spirituality.

Life Between Lives- For much of the Biblical description of death to make sense, reincarnation must be one alternative. That being said many souls sleep after death to be awakened at what has been called the final days or the time of resurrection.

Memory in Heaven- Certainly, our head memories are no still with us in heaven, but the collection of feeling, events, and memories collected by the Heart brain and soul accompany us after death. Sometimes reincarnated children remember some of these items up until about age 9, but soon lose those memories as they are clouded by head-brain memories push them aside.

God Exists- Unfortunately, the universe cannot exist alone. A controller keeps everything from slipping back into entropy and disappearing. This God is the creator of everything and from the other books, physics, biblical descriptions, and just sensing the unbelievable complexity of our Self Soul, and Spirit, we cannot possibly feel an uncontrolled expansion of life from random chemical reactions. The Heart-Brain is vastly important to our lives and could be even more important if we would just let it, but without God there would be no existence so it wouldn't make any difference.

With that, I must end this book.

About the Author

Steve Preston is a long lime author of scientific, esoteric facts. His books focus on the painful truths rather than whitewashed details that make us comfortable. If you are interested in the truth instead of comfort, please review other works by Mr. Preston as shown below. The images are some from Egypt taking the older version of taxi similar to what Moses might have used. To the right the writer is shown in the Jewish Negev desert of Israel where the Dead Sea Scrolls were found that were used by John the Baptist in his teachings.

His books include a wide assorttment of different subjects inlcuding Biblical History and proofs, the story of man's development, Ancient tecnology, New veiws of Physics and Biology, Anceint Wars, current fears and events. A partial list follows.

Development of Mankind

The First Creation of Man-book 1 History of mankind
The Second Creation of Man-book 2 History of mankind
The Creation of Adam and Eve-book 3 History of mankind
The Antediluvian War Years-book 4 History of mankind
Man After The Flood-book 5 History of mankind
A Closer Look at Ancient History-book 6 History of mankind
A New View of Modern History-book 7 History of mankind
The Twentieth Century and Beyond- Book 8 History of Mankind

Bible History, Correction, and Analysis

Abraham to Moses-First part of the Bible
Adam's First Wife-Story of Lilith
Adam to Abraham- Second Part of the Bible
Closer Look At Genesis- 200 ancient text confirm Genesis
Exploring Exodus- Reviewing the Details of "Exodus"
Errors in Understanding- Interpretations of the Bible
Expanded Genesis- Apocrypha and other Jewish texts
Exploring Genesis- Reviewing the details of "Genesis'
Incarnations of God- How often did God become Incarnated?
History Confirmed By The Bible- Science confirmation of the Bible
Moses Saved Egypt- How the Jews eliminated the Hyksos
Moses to Jesus- Third part of the Bible Series
Mysteries of the Exodus- Proofs of the Exodus
New look at the Bible- Questions in Interpretation
Old Testament Used By Jesus- Ancient Jewish texts
Understanding the New Testament-4th part of the Bible Series
Why the King James Bible Failed- Issues with KJB

Ancient Technology and Life

Anakim Gods- History of the Ancient Giant/gods

Ancient History of Flying- Ancient flying
Kingdoms Before the Flood- Pleistocene humans
Living on Venus- Venus before the Pleistocene Extinction
Martians- Ancient Life on Mars
Mysterious Pyramids- Who made the Pyramids?
Victory of the Earth- History of our Earth
Not from Space- UFOs are not from space.

Ancient and Modern War
America's Civil War Lie- Truth about the Civil War years
Behind the Tower of Babel- Story of the Bharata War
Driven Underground- Fear in the Bharata War
Four Armageddons- The 4 major wars that destroyed mankind
Six Deaths of Man- Destructions of mankind
World War Before- The Pleistocene War
World War with Heaven- The Angel and Anak War
World War Zero-The Bharata War
When Giants Ruled the Earth- History of the Titan Giants
Sex Crazed Angels- What caused the Heaven War?

Current Events and Fears
Allah' God of the Moon- Terror of Muslims
American School Disaster- fear in our country
Can We Save America? - Fear in the USA
Scythians Conquer Ireland- A History of Ireland
Fast History of MILES Training- Laser based Army training
Great American Quiz- Unusual details of American History
Make Your Own Global Warming
Truth About Phoenicia- The Evidence -First in America
Monsters are Alive- Post Pleistocene Monsters
Promote the General Welfare- Fear in USA

Our Very Odd Presidents- President review
Terror of Global Warming- Fake issue uncovered
The Antichrist- Many demonic possessed rulers
The Bad Side of Lincoln- Negative side of a great man
The Devil- Of Demons and their master
Vampires among Us- How Demons and Vampires are similar
Humans on Display- Slavery and Human Zoos

New Look at Physics

Amazing Technology- Pleistocene Technology
Anthropic Reality- We control our Reality
Consensus Science- Fake Science
Complex Earth- Truth behind Earth's development
Is Time Travel Possible? Science of Time Travel
Retiming the Earth- Eliminate of Nuclear Decay Errors
Releasing Your Consciousness- Beyond our SELF
Slip Through a Wall- How to walk through solids
Our 12-Dimensional Universe- New science of our Universe
Mystery of Photons and Light- Science of Photons
Of Heaven and Hell- scientific descriptions
Meaning of Life and Light- Detains of New Science
Vibrational Matter- New Science of Quantum Fluctuations

New Look at Biology

DNA of Our Ancestors- Tracing DNA of ancient man
God Didn't Make The Ape- New science on ape Evolution
Lizard People- Mutated People of the Bharata War
Creation and Death of Dinosaurs- Why Dinosaurs died
Races of Men- Tracing DNA of Humans

Tracing Cro-Magnon to Jesus- The third creation and mutation
Self, Soul, Spirit- Three components of Life
Self-Virtualization- New science of reality
True Happiness- Self Actualism and Beyond
Life Resonance- Unusual capabilities of men
Awaken the Departed- We can talk to the Dead
Biophotonics and Healing- How Photonics used in medicine

Printed in Great Britain
by Amazon